A Nez Perce Nature Guide
I Am of This Land
Wetes pe m'e wes

Compiled & Edited
by
Dan Landeen and Jeremy Crow

A Nez Perce Nature Guide

Conceived and produced by
Nez Perce Tribe
Environmental Restoration &
Waste Management Department
Second Printing 1997

The Nez Perce Tribe Department of Environmental Restoration and Waste Management gratefully acknowledges the U.S. Department of Energy for funding this project under cooperative agreement #DE-FC06-92 RL 12539.

We all have the responsibility to protect nature. We have this responsibility because we have the ability to destroy it. The Cayuse leader Young Chief asks, "I wonder if the ground has anything to say? I wonder if the ground is listening to what is said? I wonder if the ground would come alive and what is on it? Though I hear what the ground says… It was from me man was made. The Great Spirit, in placing men on the earth, desired them to take good care of the ground and to do each other no harm…" Young Chief listened to the earth and learned. Hopefully you will use this book to increase your appreciation and knowledge of nature and the world upon which we all live.

The Nez Perce Tribal
Executive Committee:
Samuel Penney, Chairman • Della Cree • Julia Davis • Del T. White • Tonia Garcia • Carla HighEagle • Wilfred Scott • Jamie Pinkham • Arthur Taylor, Jr.

Project Editor: Dan Landeen
Layout and Design: Jeremy Crow
Copyright © 1997 Nez Perce Tribe

All rights reserved. No part of this publication may be reproduced or transmitted in any form or by any means without the written permission of the publisher.

Western Printing
833 6th Street
Clarkston, WA 99403
(509) 751-9563
ISBN 1-891055-00-3

© 1995 Nez Perce Tribe

Printed by Western Printing, Clarkston, Washington

Printed on recycled paper

for
J. Herman Reuben
1930–1996

The Hanford Site, including the Columbia River, has a history since time immemorial as a gathering place for Indian nations to hunt, fish, trade and feast. The Nez Perce have shared and participated in these known ancient and traditional activities with other tribes where there were no fences, boundary lines or treaties.

"In the Indian world, it is known that when you take something from the environment, it must be replaced to maintain balance and order. At Hanford, what was taken away was all that was natural and the replacements were things synthetic, artificial, and in all ways "fake." Hanford has taken away our traditional foods; now we use grocery stores. It has polluted the air and contaminated the river and groundwater which affects our health and destroys the fish and animals. We must remember that the Creator gave us one Mother Earth that provides food, one air that we breathe for life, one water to nourish our body, animal, and plant life. We must continue as caretakers of the earth, or life will surely end soon."

HERMAN REUBEN, NEZ PERCE TRIBE
CHAIR, HANFORD TRIBAL SERVICE PROGRAM ADVISORY BOARD

This publication is dedicated to the memory of J. Herman Reuben who passed away in May of 1996. He was a highly respected representative for the Nez Perce Tribe; his patience and positive attitude was appreciated by all. Herman contributed much to the clean up effort at the Hanford nuclear site, particularly the preservation of cultural and natural resources. He had an appreciation for all the plants and animals that reside at the Hanford Site and always strove to learn more. Herman was also an accomplished Nez Perce storyteller and some of his coyote stories that he loved to share are included in this publication.

TABLE OF CONTENTS

FOREWORD
6

INTRODUCTION
7

SECTION ONE
Culture of the Nez Perce
Values
Stories
Nez Perce Use of Plant & Animal Resources
Hunting
Fishing
8

SECTION TWO
Historical Resource Use on the Columbia Plateau
Nez Perce History
Hanford Site
18

SECTION THREE
Hanford Wildlife
Birds
Mammals
Reptiles & Amphibians
Dangerous Creatures
25

APPENDIX
Nez Perce Treaty of 1855
Bibliography
Acknowledgements
Checklist of Animals
88

Foreword

I invite you to enjoy this adventure into the wildlife treasures hidden within the natural beauty of the lands in and around the Nez Perce Tribe's native homelands. As you will see throughout this book, there is a broad variety of species from which Nez Perce storytellers have weaved stories rich with culture and tradition. The Nez Perce have always considered that land and its creatures as essential to everyday life. Humans are considered to be only one small part of a much larger circle of life on earth. Although the risks of modern society have created significant challenges to our lifestyles, it is important to study and protect the life forms that may be even more susceptible to the environmental risks of today. The focus of this guide is the Hanford area of central Washington State. The Nez Perce have historically utilized this area in their travels to the Columbia River. I invite you to enjoy the color photographs, traditional stories, and detailed descriptions of many of the creatures that inhabit this diverse area.

Sam Penney
Chairman, Nez Perce Tribal Executive Committee

Introduction

The Inland Northwest, and specifically the Hanford nuclear site of central Washington, is an arid country some view as a wasteland. Although the land is harsh, it is home to an abundance of plant and animal life, many unique to the area. The Hanford site, long off-limits to the public, contains the region's largest undisturbed tracts of shrub-steppe habitat. It has become a haven for many plants and animals native to the region.

I Am of This Land was written to provide a glimpse of the richness of animal life found in the Inland Northwest. It provides information about the animals of the Hanford site, however the true value of the book is that it hearkens back to a time when these animals occupied the broad stretch from the Cascades in Washington and Oregon to the Bitterroots in Idaho. This was a time when the Nez Perce, along with the other Plateau tribes, lived in a traditional manner, as part of the environment.

I invite you to read *I Am of This Land* to enjoy the richness of the Nez Perce culture and wonder of the animal life found in the Inland Northwest, then use it as a reference guide for your own expeditions into this beautiful landscape.

Donna Powaukee
Nez Perce ERWM Manager

The Nez Perce Indians traditionally held a great reverence for the circle of life and all of its components. Plants, animals, even the earth itself were cherished and protected. Nez Perce stories exemplify this intimate relationship between humans and the earth. Every plant had a purpose; every animal had a name. This book discusses these animals, particularly the animals of the Hanford area of central Washington state and their significance to the Nez Perce culture. It includes not only the status of mammals, birds, reptiles, and amphibians resident at Hanford, but offers a glimpse into the Nez Perce world and other Columbia Basin tribes through their stories, language and customs. Color pictures accompany the descriptions of both dangerous and sensitive wildlife, culturally important fish, and many of the common animals found at Hanford.

Columbia River

The Nez Perce have a long history of utilizing the Mid-Columbia River Basin. The Treaty of 1855 between the Nez Perce Tribe and the United States guaranteed aboriginal fishing and gathering rights to areas in Idaho, Washington, and Oregon. Since 1855, a series of federal and state court decisions have recognized and reaffirmed the reserved treaty rights of the Nez Perce Tribe. These rights include the right to use their *usual and accustomed* fishing and gathering areas (see APPENDIX 1). This right extends to natural

> The earth is part of my body… I belong to the land out of which I came. The earth is my mother.
> —*Too-hool-hool-zute, Nez Perce*

resources in the Hanford Reach area, providing a basis for the Nez Perce Tribe/Department of Energy relationship.

The ecosystems which support fish and wildlife must remain undamaged and productive to ensure the continuation of Nez Perce culture. For this reason, the Nez Perce Tribe considers the protection of the cultural and natural resources associated with the Columbia River and the Hanford ecosystem to be of the utmost importance. The Nez Perce Tribe is especially concerned about Hanford work

Mule Deer near Hanford's White Bluffs

> Our traditional relationship with the earth was more than just reverence for the land. It was knowing that every living thing had been placed here by the Creator and that we were part of a sacred relationship…entrusted with the care and protection of our Mother Earth, we could not stand apart from our environment.
> —*Elsie Maynard (Nez Perce)*

activities and potential land-use decisions that may impact or destroy undisturbed sagebrush/steppe habitat and riparian habitat along the Columbia River.

VALUES

Traditional Nez Perce culture wove an intimate relationship between humanity and nature. In all phases of their daily lives, the Nez Perces recognized the spirits of the forces and objects around them as supernatural guardian forms which they called in a personal way their *Wyakin*. The Wyakin belief reflected a Nez Perce universe filled with individual spirits that existed in both the spirit world and the real world.

The Nez Perces identified themselves with all the natural features of the earth. The earth, in the Nez Perces' belief, was the ever nourishing mother,, as any mother provides for a child. The Nez Perces lived in a state of balance and harmony with their surroundings. They were kin to the animals and trees, to the grasses seared by the sun, to the insects on the rocks, the brooks running through snowbanks, and the rain dropping from the leaves of bushes. Everything about them, the inanimate as well as the animate, was bound like themselves to the earth and possessed a spirit, intimately joining the whole of creation.

Many ceremonies included expressions of respect for animals. The Nez Perce believed that before humans were created, animals dominated the earth and behaved like humans. It was believed that they only became mute after

> There was a time when the animals could talk and act like people, but they were still animals.
> —Alex Pinkham (Nez Perce)

COYOTE MAKES THE HUMAN BEINGS

One day, long before there were any people on the earth, a monster came down from the north. He was a huge monster and he ate everything in sight, including the chipmunks, raccoons, mice, deer, elk, and mountain lion. Coyote decided the time had come to stop the monster. Coyote went across the Snake River and tied himself to the highest peak in the Wallowa Mountains. Then he called out to the monster on the other side of the river. He challenged the monster to try and eat him. The monster charged across the river and tried as hard as he could to suck Coyote off the mountain with his breath but it was no use. Coyote's rope was too strong. This frightened the monster. He decided to make friends with Coyote and he invited Coyote to come and stay with him for a while. One day Coyote told the monster he would like to see all of the animals in the monster's belly. The monster agreed and let Coyote go in.

When he went inside, Coyote saw that all the animals were safe. He told them to get ready to escape. With his fire starter he built a huge fire in the monster's stomach. Then he took his knife and cut the monster's heart out. The monster died a great death and all the animals escaped.

Coyote said that in honor of the event he was going to create a new animal, a human being. Coyote cut the monster up in pieces and flung the pieces to the four winds. Where each piece landed, some in the north, some to the south, others to the east and west, in valleys and canyons and along the river, a tribe was born. It was in this way that all the tribes came to be. When he was finished, Fox said that no tribe had been created on the spot where they stood. Coyote was sorry he had no more parts, but then he had an idea. He washed the blood from his hands with water and sprinkled the drops on the ground. Coyote said, "Here on this ground I make the Nez Perce. They will be few in number, but they will be strong and pure." And this is how the human beings came to be.

humans were created, however they could still reveal messages or advice to humans in dreams and visions or withhold their flesh from people if they were offended. Each kind of animal represented a guardian spirit who could grant people powers and talents for success in life.

> In Nez Perce lore the coyote is a mystical being who can change himself into anything he wants. He was a being who made all the mistakes a human could make—and he had to learn.
> Ronald Pinkham (Nez Perce)

Stories

Nez Perce stories imparted basic beliefs and taught moral values. Stories helped explain the creation of the world, the origin of rituals and customs, and the meaning of natural phenomena. Stories were a way of teaching children, promoting communication, and entertainment.

They were normally told by the elders during the winter and during travel. These stories imparted upon children a moral framework and taught them such things as the habits and origins of animals, the location of food and other geographical features, how to use tools and weapons, and cultural traditions.

Stories that include Coyote make up the majority of Nez Perce stories. Because the Nez Perce believe that Coyote helped create them, he is very prominent in the stories.

Coyote is very complex in his nature and has a great capacity to survive. He represents many of the most basic human drives including lust, power and hunger for food.

Coyote is often referred to as a trickster but in many of the stories Coyote is also the victim of tricks perpetrated by others. At times it was Coyote's greed, curiosity and lack of foresight which were blamed for human hardships such as childbirth, winter, and death.

Coyote isn't the only animal featured in Nez Perce stories. Horned lizard and sculpin were considered to be weather-changers and doctors. If the horned lizard was treated disrespectfully one should expect a return of winter at root digging time. It was believed that horned lizard could cure the ill or injured by blowing short puffs of healing breath. Rattlesnakes were also considered to be powerful doctors. Crayfish were not eaten because they too were doctors. Metallic wood boring beetles were thought to help in gambling. Cicadas are

known as thirst-makers because of the hot summer days when the cicadas are making the most noise. Ticks and mosquitoes were thought to be canni-

bals, reduced at Coyote's command to pests that are with us to this day.

Stories sometimes showed ecological relationships. One story tells of Golden Eagle and his five daughters. They come in order of size: first Kestrel, then Prairie Falcon, then Red-tailed Hawk, then Ferruginous Hawk, and finally, Osprey. Nez Perce stories scarcely mention the bald eagle. "Eagle" is always the prince of raptors, Golden eagle.

Birds are important as signs. Swallows signal the return of the spring Chinook salmon, while the oriole repre- sents the rebirth of life in springtime. The canyon wren, said to attack and drive off rattlesnakes in defense of its rimrock nest, exhibits bravery in defense of home and family. Meadowlark's song gives its flesh "medicinal" power to cure speech impedi- ments, while the mourning dove's heart is eaten to make a person "quiet." Mead- owlark also speaks to Indians, but in a teasing

way.

Ravens are power- ful Indian doctors and messengers. Those with raven's power can interpret the messages they bring of portentous distant events. Generally, ravens bear news of significance over long distances without the chill import of the Great Horned Owl's tidings. Conversely, talkative crows rarely say any- thing of significance, being given over to idle gossip. The raven is a more solitary bird partial to wilderness; the crow is a gregarious camp-follower.

Feathers were collected for many reasons including aesthetic and religious purposes. Hawk feath- ers kept an arrow's flight true. The bright red-orange flight feathers of the flicker adorned dancers, and the broad tail feathers of eagles (both bald and golden) were essential to dress the corpse of

the deceased for the journey beyond. Eagle feathers have intrinsic power; if war dancers should drop an eagle feather on the dance floor the competition is stopped until a ceremony is performed to pick up the fallen feather. Eagles were seldom killed but were taken from the nest when young. The first set of feathers was plucked and a part of the second set; then the bird was released.

The excerpt above from the journals of Lewis and Clark illustrate the importance of animals in the lives of the Nez Perce.

"The *Cho-pun-nish* or Pierced nose Indian are Stout likely men, handsom women, and verry dressey in their way, the dress of the men are a White Buffalow robe or Elk Skin dressed with Beeds which are generally white, Sea Shells and the Mother of Pirl hung to their hair and on a piece of otter skin about their necks hair Ceewed in two parsels hanging forward over their Sholders, feathers, and different Coloured Paints which they find in their Countrey Generally white, Green and light Blue. Some few were a Shirt of Dressed Skins and long legins and Mockersons Painted, which appears to be their winters dress, with a plat of twisted grass about their Necks. The women dress in a Shirt of Ibex or Goat Skins which reach quite down to their anckles with a girdle, their heads are not ornemented, their Shirts are ornemented with quilled Brass, Small peces of Brass Cut into different forms, Beeds, Shells and curious bones."
—Lewis & Clark journal

Mariposa lily

Desert parsley

Nez Perce Use of Plant and Animal Resources

Hunting and fishing activities were very important components historically in the lives of the Nez Perce and other Mid-Columbian tribes. Among the Nez Perce, hunting contributed 10-25 percent of the diet and fishing contributed 30-40 percent. Gathering activities made up the remainder of the diet. Many species of plants that occur at Hanford have a long history of traditional use by the people who lived there. Some of these plants include wapato, several species of desert parsley, balsamroot, yampah, clusterlily, and mariposa lily. All of these plants were harvested with digging sticks that were usually made from bones or antlers from deer or elk. Women harvested these roots first on sunny south-facing slopes and later during the season on north-facing slopes. The Nez Perce made seasonal migrations throughout their inland northwest territory in order to take advantage of various plant and animal resources.

Hunting

Hunting skills were very important

and highly valued. Animals provided food, clothing, and ceremonial objects. A boy's first kill was a rite of passage to manhood and was also a traditional prerequisite to marriage. Animals hunted in the Mid-Columbia Basin included bison, pronghorn antelope, mule deer, elk, bighorn sheep, rodents, and birds. Bison and pronghorn antelope are two species that have been

> The buffalo are gone—will the salmon be next?
> —Julia Davis (Nez Perce)

reported from eastern Washington sites but do not presently occur at the Hanford Site. Bison were apparently present into the late prehistoric period in the Columbia Plateau but the ecology of the species near the margins of its natural distribution is poorly known. Bison remains have been recovered from the Hanford Site. Pronghorn antelope remains were recovered in large numbers near the mouth of the Snake River on Strawberry Island. Strawberry Island yielded a total of 51,322 individual bones, 92 percent which consisted of pronghorn bones.

> This war whistle which helped me in dangerous places is made from the wing bone of the crane. Spirits guided me in its making... Only in battle or other dangerous places did I sound this whistle. Not at any other time was it to be sounded.
> —Yellow Wolf, Nez Perce

Nez Perce hunters used bows and arrows before guns were known. Wasting game of any kind was a serious offense against both the animals and nature and was believed to be punished with sickness or bad luck in hunting. The Nez Perce used every part of the animals they killed, partly out of respect to the animal and partly because of the utility of the products which could be obtained from them. For example, deer, elk, and buffalo meat was not only consumed at the time of the kill, but was made into jerky for the winter months. Women relied on animal's brains to clean and soften hides in order to make moccasins and buckskin clothing. Hunters used hides to conceal themselves when hunting animals.. Dancers wore dance rattles made from mule deer hooves, and mule deer and elk teeth adorned many articles of Nez Perce clothing. Bone and antler provided the raw material for many tools and implements. Nez Perce craftsman relied on bone to craft arrow points, fish spears hooks, and arrow flakers. Bone awls were used for puncturing, braiding, weaving, and sewing. The long bones of Sandhill cranes and eagles created whistles. Gambling pieces were also carved from bone. Nez Perce hunters sometimes used elk antler bows; craftsman felled trees, split wood, and hollowed out canoes with elk and deer antler wedges; and cooks relied on mountain sheep horn bowls and spoons.

Nez Perce hunters perfected a variety of weapons, but the best was a distinctive bow which eventually won fame among other Indians who eagerly sought it in intertribal trade. The Nez Perces made the bow, which was about three feet long, from a section of the curled horn of a mountain sheep. After straightening it by a patient process of steaming and stretching, they backed it with deer sinew attached by a glue made from the scraped skin of a salmon, deer or elk hooves, or the boiled and dried blood of a sturgeon. The finished bow was handsome and powerful, and with it the Nez Perces could shoot arrows, as long as the bow itself, clear through the bodies of running animals. Sometimes, the Nez Perce made poison arrows in preparation for war by provoking captured rattlesnakes into striking at pieces of liver which they then smeared onto arrowheads.

The Nez Perce also utilized many other animals. Weasels were highly esteemed because of their ability to change their pelt color and the effectiveness with which they killed their prey—including many larger than themselves. Weasel and otter skins were commonly used as part of the traditional dress. Weasel skins were more often used for decorative and ceremonial purposes, whereas, otter skins were made into gloves and leggings. The yellow-bellied marmot was hunted near summer fishing grounds. Otters and beaver were also considered delicacies and beaver teeth were made into necklaces. Porcupine quills were softened and utilized for decorative purposes.

Townsend ground squirrels and the Washington ground squirrel were also utilized for food. Today the Townsend ground squirrel is the only ground squirrel that resides at the Hanford Site. Sometimes streams were diverted to flood Townsend ground squirrel colonies. Jackrabbits and cottontails were netted in sagebrush flats in communal hunts using long hemp nets and the rabbit fur was used for winter vests and socks. Trapping was done for the taking of beaver, muskrat, and otter. Otter skins were used for decorative and symbolic hair braiding; and beaver musk glands were used as an aphrodisiac and love charm. Many of the smaller animals were snared with simple running

> White Spirit God make these nice land, tree, animals, birds, fish, much more…and roots…for Indian people and he make present for Indian people to live here good.
> —George Peopeo (Nez Perce)

nooses. Magpies were captured at the nest site with a noose at the entrance. Rabbits and deer were hunted in drives. Coyotes, wolves, badgers, and bobcats were caught in deadfall traps. Sage grouse were highly prized and shot with bows and arrows. Sage grouse no longer exist at the Hanford Site.

The Nez Perce recognized about 70 of the bird species found in the Mid-Columbia region. Twenty-one

duck species share a single generic Sahaptian name and many were hunted, such as the Canada goose on islands in the Columbia River. Eggs of some species of waterfowl were also collected. Mallard ducks and the common merganser were among the most frequently hunted waterfowl. Tundra and trumpeter

> The tribes always treated water as a medicine because it nourished the life of the earth, flushing poisons out of humans, other creatures and the land. We knew that to be productive, water must be kept pure. When water is kept cold and clean, it takes care of the salmon.
> —Levi Holt, Nez Perce

swans that wintered along the Columbia River provided additional winter food.

Fishing

The majority of the Nez Perce diet was fish. For this reason, Nez Perce fishermen were highly skilled. Their most prized fish was the chinook salmon. The favored fishing sites occupied a 10 mile stretch of the Columbia River between Celilo Falls and the Dalles. Salmon were caught with nets, seines, gaffs, and weirs. Access to

these sites was controlled by local villagers. A salmon chief had authority to open and close fishing seasons. Fishing was closed at night until late the following morning to allow some salmon to escape and fishing was not allowed on funeral days. Salmon were eaten fresh, or dried for winter consumption. Broiled salmon were always cut lengthwise into three slabs, one slab from each side and one containing the spine. Salmon were also used to make pemmican.

> Put the fish back in the rivers and protect the rivers that support them
> —Sam Penney (Nez Perce)

At the present time, the future of chinook salmon and other fishes in the Columbia River

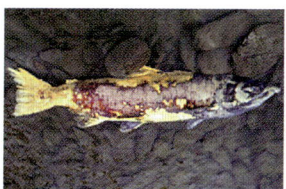

Coyote and the Five Swallow Sisters

Five swallow sisters had blocked the salmon migration with a dam. Coyote heard of this while traveling down the river and decided to do something about it. Coyote was too well known to approach the sisters directly so he turned himself into a baby strapped on a cradle board and set himself adrift in the river. Soon he lodged up against the sister's dam where he was discovered by the youngest of the five. "Oh, look at the poor baby," they all cried, and took him home with them. The next day they all set off to go root digging in the hills leaving the baby.

As soon as they were out of sight, Coyote turned himself back into his real self and set to work. He made himself five digging sticks and five oak-burl mortars. He then attacked the dam with the first digging stick. For four days he worked, each day with a new digging stick, stopping only as the sisters approached the village. On the fifth day, as the sisters were digging roots, the youngest's digging stick broke. This being an ill omen, they all ran back to the village suspecting that something was amiss with the baby.

They found Coyote hard at work with his last digging stick, breaking up their dam. Coyote quickly put on the first oak mortar as a helmet to protect himself from the sister's attack. They broke first one, then the second, third and fourth. But Coyote was nearly through the dam, which gave way in a flood just as the fifth mortar helmet broke. The sisters were defeated and the salmon swam up to join the people. Thereafter, the swallow sisters mud nests must signal the return of Chinook salmon each spring.

25-40 percent in young salmon which eat insects as well as the carcasses. In another study conducted on the Olympic Peninsula, researchers tossed 900 salmon carcasses into eight streams. Twenty different mammal species and 23 bird species fed on these carcasses including shrews, squirrels, beavers, elk, deer, sparrows, woodpeckers, chickadees, and wrens. The loss of salmon have also been implicated in the decline of resident killer whales which depend on the salmon for 80-90 percent of their diet.

Suckers caught in weirs and by snagging were highly valued as a food fish. Fishermen caught them in early spring, two months before the salmon showed up. The bones of a sucker's skull are not fully fused; when cooked, these bones fall apart, providing "storytellers" opportunities for recounting the

system is in jeopardy. Factors such as off-shore fishing, pollution, and the construction of dams on the Snake and Columbia Rivers have exacted a heavy toll. The loss of the salmon would have great cultural and ecological impacts. Most tribes in the Pacific Northwest, including the Nez Perce, have a sacred relationship with the salmon. Recent studies in Washington indicate the importance of the salmon cycle to the entire ecosystem. In a study done along Grizzly Creek, a tributary of the Snoqualmie River in western Washington, nitrogen and carbon were traced from dead salmon into insects, riparian vegetation, and other fish. Eighteen percent of the nitrogen in riparian vegetation came from salmon, 25-35 percent of the nitrogen and carbon in insects, and

source of these bones. Some of the bones were named Grizzly's earring, Raven's socks, Stellar's jay, Softbasket woman monster, and Cricket packing her child.

Eels or lampreys were considered a delicacy by all the tribes. They were roasted over a fire or dried. Today they are difficult to find on the Columbia River. Other fish caught included shiners, trout, and mountain whitefish. Fresh water mussels which can grow to a large size were also eaten in historic times. Archeologists have found deep shell middens at many of the village sites along the Hanford Reach.

With the coming of the white man the traditional lifestyle of

fishing and hunting was seriously jeopardized. The following account by Sam Fisher who was the last independent Palouse Indian who lived next to the Columbia River exemplified this loss. Sam sold some beaver pelts and the game warden took him to court. In the courtroom he told the following story:

"My father and his father's father lived here. They and my people lived here. They were here a long time ago and for a long time. When they grew old, they died and returned to their Mother.

"A long time ago the Palouses killed deer and beaver. We fished in the river. That was how we got food and clothing. There was no one to tell the Indian what he could do and what he could not do.

"The salmon came every year as did deer and beaver, the way the Watcher intended it should be. And we used them as they were intended to be used.

"But when the big herds of cattle wandered over the Palouse Hills and when the

> When you gaff, you just don't put your pole in the water and pull something out. You feel for the fish. It's a touch to feel for them.
> —Larry Greene Sr. (Nez Perce)

sheep came later, they ate all the grass —even to the very doors of our lodges—so the deer no longer came.

"The white man built dams that choked the rivers; only a few salmon could come.

"Now I am accused of killing beaver. I did.

"You say it is not the law to kill beaver. I ate the flesh because I needed food. So did my family. I sold the skins because I needed money to buy clothes for my wife and children. When I did this, I stole nothing that belonged to the white man."

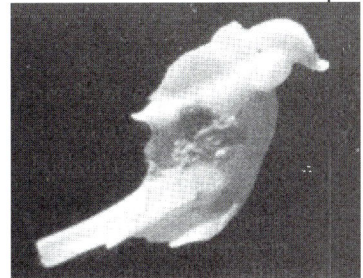

Sucker skull bone called Stellar's jay

History of Resource Use: Columbia Plateau

Evidence for human occupation in the region encompassing the Hanford Site dates as far back as 11,000 years and possibly beyond. The earliest inhabitants of the region hunted large mammals and foraged opportunistically. As a seasonal round of resource exploitation gradually emerged, small bands of people lived in river shelters (e.g. caves, rockshelters), stored some food, and began using fish and shellfish. This lifestyle evolved into a greater dependence on river resources. Pithouses (semi-subterranean dwellings) characterized settlement during the period from 4,000 to 2,000 years ago as the people continued to follow a modified seasonal round, storing food and living in small groups, and using settlements as base camps. River and major creek settings began to form the focus of settlement and activities, although groups would travel to other areas to utilize other resources. About 3,000 years ago, the lifestyle became more permanent along the rivers of the region.

The people who occupied the mid Columbia Basin region did not become well defined until comparatively recent times. Because of common defense systems, trade, and the gathering of food and unions between the different bands, many of the tribes shared and continue to share common beliefs and traditions. The unification of peoples into tribes in the political sense occurred only after the arrival of the white man.

Nez Perce History

The Nez Perce, who call themselves *Nimiipu* [the real people], first came into contact with Euro-Americans in 1805 with members of the Lewis and Clark Expedition as they descended onto the Weippe meadow above the Clearwater River. The Tribe's reputation for friendliness began with this visit. Traditionally the Nez Perces lived in scattered villages and maintained few political institutions. Tribal life evolved around small,

> The earth was created by the assistance of the sun, and it should be left as it was... The country was made without lines of demarcation, and it is no man's business to divide it... The earth and myself are of one mind. The measure of the land and the measure of our bodies are the same... Understand me fully with reference to my affection for the land. I never said the land was mine to do with as I choose. The one who has the right to dispose of it is the one who has created it.
> —Chief Joseph

semipermanent villages that lay along the shores of the major streams and creeks. The village size varied upon bands ranging from 10 to 75 members with over 300 known Nez Perce village sites in Oregon, Idaho, and Washington encompassing over 13.5 million acres.

Because their survival required that the bands move in an annual gathering cycle,

there were no permanent sites and very little extended political organization beyond the band headmen and peace leaders who ensured that the women, elderly, and children were provided for. The tribal identity was derived from the commonality of language, land, family, and religion. Euro-American contact brought the horse in the 1730's and many new trade and warfare items.

The coming of the horse dramatically changed the Nez Perce lifestyle. With greater mobility they traveled more often to the buffalo country in Montana as well as on trade missions to the Columbia Basin region. The longhouses were still utilized in the winter months, but families adapted the portable tepee style dwellings. The Nez Perce, along with the Cayuse tribe, were the only known tribes to selectively breed horses to improve their stock. This was accomplished by culling horses that exhibited inferior traits. This practice added to the wealth and reputation of the Nez Perce in terms of trade goods and territory.

> If the white man wants to live in peace with the Indian he can live in peace.. There need be no trouble. Treat all men alike. Give them all the same law. Give them all an even chance to live and grow. All men were made by the same Great Spirit Chief. They are all brothers. The earth is the mother of all people, and all people should have equal rights upon it. You might as well expect the rivers to run backward as that any man who was born free should be contented penned up and denied liberty to go where he pleases.
>
> —Chief Joseph

The trade practice was increased not only within bands of the Nez Perce, but with outside tribes, including the Yakama, Umatilla, Walla Walla, Palouse, Blackfeet, Crow, and Sioux.

Shortly after the introduction of the horse, the first known Euro-Americans introduced epidemics that brought devastation and death to all tribes of the Plateau. The first documented smallpox outbreak among the

Plateau Indians was in the 1770's and another outbreak in 1801 resulted in a population reduction of around 45 percent. Although no reliable early census data are available, the Nez Perce reportedly numbered around 6,000 at the time of Lewis and Clark.

In 1836, the Nez Perces began a new relationship with Euro-American influence that produced a permanent change that still affects the tribe today. During this time a group of Presbyterian ministers arrived in Idaho and settled in the heart of Nez Perce country. Thus began the fragmentation of the Nez Perces based on spiritual and material survival needs. Those Nez Perces who adopted the Christian religion found advances could be made in terms of material wealth and land acquisition. Those Nez Perces who clung to the Dreamer religion soon realized that this division would tear the Tribe apart for many generations.

At mid-century, the Indian Office began moving Indians in the Northwest onto reservations to separate them from the growing number of white settlers as well as for religious differences. The treaty of 1855 resulted in 7.5 million acres set aside for the Nez Perce. It also required recognition of the American government and the imposition of an Office of Principal Chief which was not acceptable to many bands of the Nez Perce. The official recognition of Head Chief was assumed by Chief Lawyer of Kamiah, who was friendly with the whites and also a Christian convert. When gold was discovered on the Nez Perce reservation in 1860, his band was helpful in bringing supplies to the miners. In 1863 another treaty was signed which

> "My son," said Old Joseph, "my body is returning to my mother earth, and my spirit is going very soon to see the Great Spirit Chief. When I am gone, think of your country. You are the chief of these people. They look to you to guide them. Always remember that your father never sold his country. You must stop your ears whenever you are asked to sign a treaty selling your home. A few years more, and white men will be all around you. They have their eyes on this land. My son, never forget my dying words. This country holds your father's body. Never sell the bones of your father and your mother."
>
> To that Chief Joseph added, "I buried him in that beautiful valley of winding waters. I love that land more than the rest of the world. A man who would not love his father's grave is worse then a wild animal.
> —Chief Joseph

> The white men were many and we could not hold our own with them. We were like deer. They were like grizzly bears. We had a small country. Their country was large. We were contented to let things remain as the Great Spirit made them. They were not, and would change the rivers if they did not suit them.
> —Chief Joseph

greatly divided the Tribe. It reduced the acreage to 750,000, abandoning claims to lands in Oregon, Washington, and parts of Idaho. The land occupied by Chief Lawyer and his band was not ceded to the United States, however; the lands where the "dreamer" bands still resided were subsequently ceded. In 1887 the Dawes Allotment Act was passed resulting in 500,000 of those acres to be opened for white settlement. Today, only 13 percent of the Nez Perce reservation is still owned by tribal members.

In 1891 the Bureau of Indian Affairs agency census for the Nez Perce was 1,700 and the 1892 census was 1,828 which was a significant decline since the arrival of Lewis and Clark. Today there are currently 3,200 enrolled Nez Perce, of which 2,450 reside on or near the Nez Perce reservation.

Hanford Site

The Columbia Basin was not an easy land to live on and was not very attractive to the early 19th century non-Indians who explored the area for routes, furs, and economic prospects. The area had very little wood for fuel and fencing and the dry arid climate made it difficult for agriculture. The discovery of gold in lands adjacent to the Columbia Basin resulted in the town of White Bluffs on the Columbia River which is located on the Hanford Site. White Bluffs was the area for crossing the Columbia River and became a favorite place to ford cattle and supplies. This lead to the establishment of ranches throughout the area and shortly thereafter the railroad companies founded the towns of Pasco and Kennewick. In the early 1900's the federal government funded irrigation projects in the Columbia Basin and settlement of the area began in earnest.

In the 1940's the Hanford Site which initially consisted of 640 square miles in south-central Washington was selected for the location of the first, full-scale plutonium production

> My people, what have you done? While I was gone you have sold my country. I have come home and there is not left me a place on which to pitch my lodge.
> Chief Looking Glass (Nez Perce)

plants in the world. Native peoples and settlers were no longer allowed to reside at the site and were given 30 days to leave. The Hanford Site, today consists of 570 square miles and has been administered by the Department of Energy or its predecessors as a national security area since 1943. Most of the nuclear reactors along the Columbia River were shut down in 1970. The last reactor was deactivated in 1987. The mission of the Hanford Site today is to cleanup the contamination that has resulted from nuclear waste management operations.

The Hanford Site is one of the few large areas of land in the region that has not been developed or principally used for agriculture. It is also unique in that the general public's use of the area has been very restricted. The last undammed stretch of the Columbia River known as the Hanford Reach is located at the Hanford Site and is the only remaining area on the Columbia River where chinook salmon still spawn naturally. The Hanford Reach is utilized today for hunting, fishing, and recreational activities.

The climate at the Hanford Site is probably the hottest and driest in the Pacific Northwest with an annual precipitation of six inches. The Hanford Site has been classified primarily as a shrub-steppe grassland and is composed of a variety of plant communities. Major plant species include sagebrush, bitterbrush, Sandberg's bluegrass, Indian rice grass, cheatgrass, Russian thistle, and rabbitbrush. Many of the plant species which reside at the site today have been introduced including many of the trees which grow along the river.

There are several habitat types that have been identified at Hanford which include basalt outcrops, shrub-steppe, old fields, sand dunes, and riparian areas. Basalt outcrops

> I hear what the ground says... The water says the same thing... 'Feed the Indians well.' The grass says the same thing... The ground says, 'The Great Spirit has placed me here to produce all that grows on me, trees and fruit.' The same way the ground says, 'It was from me man was made.' The Great Spirit, in placing men on earth, desired them to take good care of the ground and to do each other no harm.
> —Young Chief, Cayuse

can be found on the slopes of Rattlesnake Mountain, Saddle Mountains, Umtanum Ridge, Gable Mountain, and Gable Butte. Some of the largest undisturbed tracts of sagebrush steppe habitat in the State of Washington occur at the Hanford Site. Each year more of this habitat is lost due to wildfires and work activities which support nuclear waste operations. Many species of wildlife such as the sage sparrow are closely associated with this habitat.

Several small riparian areas occur at some of the seeps and springs on or near Rattlesnake Mountain. Larger riparian and wetland areas occur along the Columbia River and provide food and cover for several species of wildlife. Many established sand dunes can be found between the 300 Area and the old Hanford townsite.

These dunes provide unique habitats for many species of plants and animals. There are many areas at the Hanford Site, especially along the Columbia River that were used for agriculture. These old fields are dominated by invader species such as tumbleweeds, cheatgrass, and mustards. Species such as the long-billed curlew nest in these areas.

SUCKER AND WHITEFISH

Once in a certain place there lived many different beings, and there were crawling beings and four-footed animals. Among these, Sucker and Whitefish were good friends.

One day a child got married, and the time came for them to go wedding trading. They took almost everything from the house when they went trading. When they got there, there was food made for them. Sucker took a half-burned log with charcoal and used it as a spoon to sip the porridge. At that instant his mouth thickened and bulged out from the great heat of the porridge and spoon. While eating the soup he said, "It takes five packloads to cook me—not just a little, but five whole packloads."

Whitefish, on the other hand, drank the soup with a straw. As he ate happily, he said, "I can be cooked with anything—a straw or anything that will burn, or even with a small piece of wood. I will cook in a short time."

When they finished eating, they left, and went back home.

This is why the sucker fish has a thick, turned-out mouth. On the other hand, whitefish has just a small pointed mouth, because he ate the soup with a straw.

Whenever you have whitefish, it will cook with just a small amount of fire. However with the sucker, you must cook it a lot and use lots of fire. Even then it barely cooks. These two have been this way ever since that time.

The earth is our Mother; we must not wound her...
—Smohalla, Wanapum

A common misconception about the Hanford Site by the public and many Hanford site workers is that the Hanford Site has a low biodiversity of plants and animals. Many people are not aware of the richness of the Hanford flora and fauna and do not realize that many species of wildlife designated as sensitive species by the state and/or federal governments reside at the site. Many of these wildlife species are also important in Nez Perce culture and traditions.

Wildlife species in this book (birds, mammals, reptiles, and amphibians) were rated as to abundance and seasonal occurrence at the Hanford Site as defined in TABLE 1. These designations are found in parentheses throughout the text when individual species are discussed. Species that are classified as sensitive species by the state and/or federal governments are also designated in the text. Definitions of the federal classification system are found in TABLE 2, while those for the state of Washington are found in TABLE 3.

TABLE 1.
Abundance and Seasonal Occurrence classifications used in this book

Abundance:

C	COMMON	often seen or heard in appropriate habitat
U	UNCOMMON	usually present but not always seen or heard
R	RARE	present only in small numbers, seldom seen or heard
A	ACCIDENTAL	documented once or twice, out of normal range

Seasonal Occurrence:

R	RESIDENT	present all year but may vary seasonally
S	SUMMER	visitor (includes spring and fall)
W	WINTER	visitor (includes spring and fall)
M	MIGRANT	

TABLE 2.
Federal Classifications of Sensitive Species

FEDERAL

E	Federal Endangered. A species in danger of extinction throughout all or a significant portion of its range.
T	Federal Threatened. A species which is likely to become endangered within the foreseeable future.
C_1	Candidate taxa for which enough substantive information is available to support listing as threatened or endangered by the federal government.

TABLE 3.
State of Washington Classifications of Sensitive Species

E   ENDANGERED. Species that are in danger of becoming extinct within the near future if factors contributing to their decline continue.

T   THREATENED. Species that are likely to become endangered within the near future if factors contributing to their population decline or habitat degradation continue.

S   SENSITIVE. Species that are vulnerable or declining, and could become endangered or threatened without active management or removal of threats.

C   CANDIDATE. Wildlife species native to the state of Washington that the Department of Fish and Wildlife will review for possible listing as endangered, threatened or sensitive.

M   MONITORED. Wildlife species native to Washington that are of special interest because: 1) they were at one time classified as endangered, threatened, or sensitive; 2) they require habitat that has limited availability during some portion of their life cycle; 3) they are indicators of environmental quality; 4) further field investigations are required to determine their population status; 5) there are unresolved taxonomic problems which may bear upon their status classification; 6) they may be competing with and impacting other species of concern; or 7) they have significant popular appeal.

Mae Taylor, Nez Perce Tribal elder gathers spring-time roots

It is morning. We are alive, so thanks be
   —Nez Perce

Birds

Birds are an important component of the Hanford ecosystem. Many of the bird species at Hanford are associated with the shrub-steppe habitat. Approximately two hundred and fifty bird species have been observed at the Hanford Site consisting of 16 orders and 38 families. Of these birds, approximately 49 are common and 47 are accidentals. One-hundred and seven of these species are listed as listed as rare or accidental and about 95 species are designated as migrants. Some of the more notable accidental species observed at the Hanford Site include the red-throated loon, mountain plover, black phoebe, red phalarope, band-tailed pigeon, Clark's nutcracker, Philadelphia vireo, rose-breasted grosbeak, rusty blackbird, gyrfalcon, long-tailed jaeger, and black-legged kittiwake. Thirty-six species of birds are listed as sensitive species. Birds such as the scaled quail and sage grouse no longer exist at the site. Scaled quail were introduced into the Columbia Basin in the 1950's and were only observed once at the Hanford Site. Sage grouse have not been seen for several years but they do reside at the Yakama Firing Center to the west of Hanford.

loons

Gaviiformes

Common loon

This order is represented by one family (GAVIIDAE) and three species; the Pacific loon (Aw), red-throated loon (Am), and common loon (Rr). The common loon is a state candidate species. An adult female common loon and one young was observed in 1978 at White Bluffs slough and in 1993 and 1996 north of Richland.

Podicipediformes

This order is represented by one family (PODICIPEDIDAE) and six species of grebes. The pied-billed grebe (Cr) and Western grebe (Ur) are the only resident species which can be observed in the Columbia River sloughs. The horned grebe (Uw), red-necked grebe (Aw), western grebe and Clark's grebe (Rw) are all state monitor species. Eared grebes (Um) are seldom observed at Hanford.

grebes

Western Grebes

Horned grebe

Eared grebe

pelicans & cormorants

American white pelican

Pelicaniformes

This order is represented by the families PELECANIDAE and PHALACROCORACIDAE and two species: the American white pelican (CR) which is a state endangered species and the double crested cormorant (CR). It has only been in the last 10-15 years that large numbers of pelicans have been observed on the Columbia River. Over 300 white pelicans have been observed at the Hanford Reach where they feed primarily on carp and suckers. White pelicans have not yet been observed nesting at the Hanford Site. Double-crested cormorants, which are also increasing in numbers, also eat fish and, like the pelicans, there are no nesting records for the site. Both species are becoming more common.

Double crested cormorants

herons & egrets

Ciconiformes

This order is represented by the family ARDEIDAE and five species. The Great blue heron (CR), great egret (RS), and black-crowned night heron (UR) are all state monitor species. The great blue heron and black-crowned night heron nest in large trees in colonies along the Columbia River near White Bluffs and the Hanford townsite. Herons, egrets, and bitterns feed on insects, fish, amphibians, and small mammals. Infrequent observations have also documented nesting by the great egret among the Great blue heron colonies. Great egrets seem to be increasing in recent years throughout all of Eastern Washington. The American bittern (Rs) and snowy egret (AM) have only been observed on a few occasions.

(Top) American bittern, (Middle clockwise) Great egret, Black-crowned night heron, snowy egret, Great blue heron, (Bottom left), Great blue heron

Coyote Visits Elk and Fish Hawk

Once upon a time, Coyote was living with his wife, Mouse. They had a child named Tse-tsa-khee. One day, Coyote said, "Let's go and visit the Elk family. He and his wife are living near by." So they all left.

When they arrived, Elk knew that Coyote was hungry and wanted meat. He said, "Let me cook meat for you."

"Yes, all right," said Coyote. Then Elk went to his wife and cut off his wife's sleeve. He boiled the sleeve because he had hot rocks ready.

Coyote wondered, "What's up? Does he expect me to eat that?" When the sleeve was cooked, Elk took it out, and it had turned into a wonderful stew. Coyote ate a great deal. He had never tasted such a good stew.

Then Coyote said to Elk, "You visit me, too."

"Yes, I will come to see you some day," Elk said.

Sometime later, Elk went to see Coyote. There, poor Coyote and his wife were living very poorly. But Coyote acted bravely, saying, "Come in. Let us serve you something." Then Coyote took an agate knife and cut the sleeve of his wife's dress, and cooked it. But when he took the sleeve out of the pot, it was the same old buckskin, and no meat at all. Elk felt sorry for Coyote.

Then Elk went back home again and Coyote said, "Now I am going to see Fish Hawk."

When he arrived, Fish Hawk was happy, saying, "Oh, we're glad to see you, Coyote. Surely you came just to see us."

"Yes," Coyote answered.

Fish Hawk knew that Coyote was always hungry, so he went and brought lots of fish, which he quickly cooked. After Coyote was finished eating, he told Coyote to take home all the fish that were left over. Coyote said, "You should visit me, too."

"Yes I will sometime," replied Fish Hawk.

One day, Fish Hawk came to visit Coyote. "It's good to see you," Coyote said, and ran down and climbed a tree. He saw that there was a hole in the ice. Coyote said, "Let me stick a twig in there to catch a fish with." But poor Coyote fell in himself and hit his head on the ice. He knocked himself out, but he soon recovered.

Fish Hawk felt sorry for him, because he knew what Coyote was trying to do. Fish Hawk said, "Wait here now, Coyote. I will go and catch fish." Then Fish Hawk went away for quite a while. When he came back, he had caught fish with a forked twig.

"Thank you," Coyote said. "I guess I just don't have the talent or means that others have to fill my needs." Too many times we try to copy the talents of others or pretend to be something that we can't be.

swans, geese & ducks

Canada geese

Anseriformes

This order is represented by 28 species in the family ANATIDAE. Several species of waterfowl like the green-winged teal (Us) utilize the Columbia River on a seasonal basis and several species such as the Canada goose (CR), mallard (CR), northern shoveler (CR), cinnamon teal (Us), gadwall (Uw), redhead (Cw), and ruddy duck (UR) nest at the Hanford Site. Canada geese nest on many of the islands in the Columbia River. Even though they nest on the islands, the geese are still susceptible to predation by coyotes. The Aleutian Canada goose (AM) and Trumpeter swan (AM) are

sensitive species that are rarely observed; however trumpeter swan observations have been increasing in recent years. Tundra swans (Rw) are observed during the spring and fall migration. Large flocks of common mergansers (Cw) also occur at the site in the winter. Several other species including the snow goose (Rw), northern pintail (Cw), Eurasian wigeon (Rw), American wigeon (Cw), canvasback (Uw), ring-necked duck (Uw), lesser scaup (Cw), greater scaup (Rw), oldsquaw (Aw),

Clockwise: Trumpeter swans, common goldeneyes, male ruddy duck, bufflehead.

common goldeneye (Uw), Barrow's goldeneye (Rw), bufflehead (Cw), and hooded merganser (Rw) can be observed in the winter. Other species rarely observed include the greater white-fronted goose (Rm), brant (Am), and red-breasted merganser (Rw).

Page 34 clockwise: American wigeon in flight, male hooded merganser, male northern shoveler. Photo to right: Mallard nest.

birds of prey

American bald eagle

Falconiformes

This order is represented by THREE families, CATHARTIDAE (vultures), ACCIPITRIDAE (eagles, hawks), and FALCONIDAE (falcons) and 17 species. Twelve of these species are designated as sensitive by the state and/or federal governments and only the northern harrier (CR), American kestrel (CR), red-tailed hawk (CR), and prairie falcon (UR) are considered year round residents. These birds prey upon insects, lizards, small mammals, birds, and snakes. Northern harriers construct ground nests at the Hanford Site and prey mainly on small birds and mammals and have been the subject of ecological studies by Battelle Northwest Laboratory at the site. American kestrels are robin-sized falcons that prefer to nest in tree cavities at the Hanford Site. They are often observed hovering in search of prey. Red-tailed hawks are the most common hawks at the Hanford Site with several pairs nesting on the site. Prairie falcons are large falcons that nest at Gable Mountain and at the cliffs on Umtanum Ridge.

Swainson's hawks (US) are one of the few species that have probably benefited from the activities at the Hanford Site. Approximately 20 pairs of these birds nest in many of the trees that still remain at the abandoned army bunker sites which were manned in the 1940's and 1950's. Ferruginous hawks (US) are large reddish-brown raptors with white underparts and a pale or white tail with no bands. Approximately 10-12 pairs of these hawks, which are classified as a sensitive species, nest at the Hanford Site, many of them in high tension electric

Prairie falcon

transmission towers. Golden eagles (Uw) are large, dark-brown birds with golden feathers on the crown of the head and nape of the neck. Most golden eagles at the Hanford Site are observed during the winter.

The turkey vulture (Am), northern goshawk (Rw), sharp-shinned hawk (Rw), gyrfalcon (Aw), Cooper's hawk (Rw), osprey (Um), merlin (Rw), peregrine falcon (Rm), and rough-legged hawk (Rw), are not often observed and many are migrants. Ospreys are fish-eating birds occasionally observed in the spring migration period along the Columbia River. Ospreys have not been observed nesting at the Hanford Site, but there have been active nests for several years along the Yakama River near Toppenish, Washington. There has also been one active nest in 1995 and 1996 across the Columbia River from Howard Amon Park in Richland. Ospreys have also been observed at some of the Hanford ponds that were made as a result of nuclear waste management operations. Peregrine falcons (Rm), which are listed as a federal and state sensitive species, migrate through the Hanford Site during spring and fall. They are about the same size as a prairie falcon with a black crown and nape and have a black wedge which extends below the eye forming a distinctive helmet. Peregrines feed almost exclusively on other birds. Peregrine falcons have been observed near the Columbia River north of Richland on several occasions.

The bald eagle (Cw), is the only wildlife species at Hanford that is listed as a federally threatened species. Approximately 40-60 bald eagles reside at the Hanford Site between November and March. They utilize the trees along the Hanford Reach of the Columbia River for perching and roosting and feed primarily on chinook salmon carcasses and mallard ducks. Bald eagles do not nest at the Hanford Site but in recent years eagles have started nest building activities. None of these attempts have resulted in successful nesting.

Osprey

(Clockwise) Peregrine falcon, Ferruginous hawk, American kestrel.

grouse & quail

Hungarian partridge

Galliformes

This order, consisting of the partridges, pheasants, grouse, and quail, is represented by the family PHASIANIDAE. Nesting by the gray partridge (RR), ring-necked pheasant (UR), chukar (UR), and California quail (UR) has been documented at the Hanford Site. Chukars are found on basalt outcrops at Rattlesnake and Gable Mountains and Umtanum Ridge. Pheasants and California quail are found in the riparian zone along the Columbia River. Gray partridges have been observed on Rattlesnake Mountain and in shrub steppe on the western boundaries of the site. The northern bobwhite (RR), which was introduced into the Yakima valley several years ago, has only been observed once.

Coyote and the Grouse

Coyote was on his way up the river, and he was getting tired from walking. He was hungry, too, for he had traveled quite a distance. He came upon a lodge where the grouse family was living, and he stopped and went inside. He found that there were only children there, and he asked those little grouse children, "Where are your parents?"

They chirped back at him, *"Wiwa."*

Then Coyote asked again, "Where are your parents?"

"Wiwa," they chirped.

Again Coyote answered, "If you children don't tell me where your parents are, I'm going to kill you all." But the children only

Sage grouse

chirped, "Wiwa" again.

So Coyote built a big fire inside the lodge and he said to the children, "Come here." They all went close to him, and he told them, "I'm going to ask you once more where your parents are. If you don't tell me, I'm going to kill all of you and eat you."

"Wiwa," they chirped.

This made Coyote very angry because they didn't answer his question. So he grabbed every one of the little children and threw them into the fire, and they all burned to death except one. That one was the youngest of the family. He got away by running through the fire and out the other side, where he hid himself in some straw. But Coyote didn't see that one of the little children had gotten away. He thought he had killed them all. The children were all cooked, so Coyote made a nice meal out of it.

After a time, the mother and father Grouse came home. They looked about the lodge, but couldn't find any of the children they had left there in the morning. Then the youngest one crawled out from his hiding place. His parents asked, "Where are the rest of the children?"

"Coyote came and killed them all by throwing them into the fire," he answered tearfully. Then he told them that Coyote had gone up the river.

This made Grouse very sad and angry, so they started out to look for Coyote. They knew that the trail he took led all the way up the river, and they knew that at one point, it led by the edge of a cliff. So right at the place, they hid themselves. Then they waited for Coyote. Suddenly, the heard him coming. He was singing, "Tsuy-tsuy liks-til ah-he ah-he. He-puht-kah-sa-yo kho-yau, pe-pa hootz pe-pa hootz, ah-he, ah-he." As he passed right beside them, one Grouse flew on one side of him, and the other Grouse flew on the other side of him. Coyote was so scared they he jumped, and fell right over the side of the cliff. He landed in the river below and was drowned.

Coyote floated down the river a bend and a half, then drifted ashore. Magpie came flying by, and he saw Coyote lying on a sandbar. So Magpie decided to stop and see if he had any eyebrow-fat. He lit on Coyote's head and began pecking at his eyebrow. Coyote woke up and said, "Yo-oh-oh. Why did you come and wake me up while I was having a good sleep? I was helping some young maidens cross the river."

Magpie said, "What? There were no maidens crossing the river. The Grouse scared you into the river and you drowned, and you floated down here."

Then Coyote got up and he remembered that Magpie was right. So he started walking up the bar, and he tapped his hip, and out came his children and started to play on the sandbar. After a while, Coyote told them, "Better come in now, before you poke each other's eyes out playing. People know that you are the sons of a brave and big chief." But he blocked the last child, Tse-tsa-khee, and said to him, "Son, tell me what has happened."

And Tse-tsa-khee told him how the Grouse had scared him off the cliff. When he finished speaking, Coyote said, "That's nothing, I knew all about that. You had better go in now."

So Coyote went on up the river. But the Grouse and their son had already left and gone farther up into the mountains, knowing that the death of their children had been revenged.

cranes & coots

Sandhill cranes

Gruiformes

The families RALLIDAE and GRUIDAE represent this order which consists of four species. The sandhill crane (CM), which migrates through the Hanford Site in large numbers, is listed as a state endangered species. Sandhill cranes utilize the islands in the Columbia River as resting places during the spring and fall migration. The American coot (CR) nests in the sloughs along the Columbia River. The sora (RS) and Virginia rail (UR), which inhabit riparian zones, are very secretive and seldom observed.

Page 40: (left) Sora, (top) American coot on floating nest constructed from rushes and cattails. Page 41:(l to r) Killdeer, Killdeer nest, American avocet

shorebirds & gulls

Charadriiformes

Bonaparte's gulls

This large order is represented by 4 families (CHARADRIIDAE, RECURVIROSTRIDAE, SCOLOPACIDAE, LARIDAE) and 38 species at the Hanford Site.

CHARADRIIDAE: *Plovers*
This family consists of the blackbellied plover (Rm), killdeer (Cr), mountain plover (Am), and semipalmated plover (Um). Killdeers which are very common construct ground nests, many times in bare or cobbled areas and will feign injury to distract potential predators.

RECURVIROSTRIDAE: *Avocets & Stilts*
This family consists of the American avocet (Us) and the black-necked stilt (Rm) which is also a state sensitive species. Avocets have nested for several years at West Lake near Gable Mountain.

shorebirds & gulls

L<small>ARIDAE</small>: *Jaegers, Gulls, & Terns*

This family is represented by 14 species which includes *Jaegers* (parasitic (A<small>M</small>), and long-tailed (A<small>M</small>)), *Gulls* (Franklin's (R<small>M</small>), Bonaparte's (U<small>M</small>), ring-billed (C<small>R</small>), California (C<small>R</small>), herring (C<small>W</small>), glaucous-winged (U<small>W</small>), and Sabine's (R<small>M</small>)), and *Terns* (Caspian (C<small>S</small>), common (R<small>M</small>), Forster's (C<small>S</small>), Arctic (A<small>M</small>), and black (R<small>M</small>)). A black-legged kittiwake (A<small>M</small>) has been observed on one occasion. Herring gulls are most often observed during the winter and are becoming more common. Colonies of nesting California gulls, ring-billed gulls, and Forster's terns have been

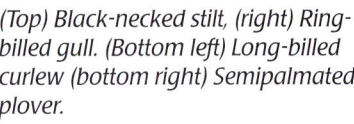

(Top) Black-necked stilt, (right) Ring-billed gull. (Bottom left) Long-billed curlew (bottom right) Semipalmated plover.

documented on many of the islands in the Hanford Reach. Terns are often observed hunting for small fish in the Columbia River sloughs, where they dive into the water in search of prey.

SCOLOPACIDAE: *Shorebirds*

This family consists of 20 species many of which are uncommon migrants (UM)—(greater yellowlegs, lesser yellowlegs, spotted sandpiper, sanderling, pectoral sandpiper, dunlin, Wilson's phalarope); rare migrants (RM)—(solitary sandpiper, Baird's sandpiper); or accidental migrants (AM)—(marbled godwit, willet (Am), sharp-tailed sandpiper, short-billed dowitcher, red phalarope). Most commonly observed species include the long-billed curlew (CS), western sandpiper (CM), least sandpiper (CM), long-billed dowitcher (CM), and common snipe (CR). The long-billed curlew is also listed as a state sensitive species which nests in many areas at the Hanford site in shortgrass prairie and sagebrush habitats.

(Top right) Common snipe, (bottom, clockwise) Wilson's phalarope, Least sandpiper, Western sandpiper, Greater yellowlegs

43

pigeons & doves

Columbiformes

(Top) Rock doves nesting in basalt cliffs, a typical nesting site for the species, (middle) Mourning dove

This order is represented by three species in the family COLUMBIDAE: the rock dove (CR), band-tailed pigeon (AM), and mourning dove (CS). The band-tailed pigeon has only been observed on one occasion at Hanford. The rock dove, or pigeon, is sometimes considered a pest, nesting in many of the abandoned reactors along the Columbia River. The

mourning dove is a game bird that nests in many areas at Hanford, especially in trees along the Columbia River.

"I learned from my friend Herman Reuben that some Indian people see the plants and animals as the old people believing we were the last, therefore youngest creatures put on earth. We are given some abilities but not all the wisdom and strength to survive in this world. So to whom do we turn to learn this wisdom and to gain strength? We turn to our elders, the older living beings, like plants and animals."

Jaime Pinkham (Nez Perce)

This order represented by the owls consists of two families (TYTONIDAE and STRIGIDAE) and 10 species. Five of these species, which include the flammulated owl (Am), western screech owl (Am), snowy owl (Rw), barred owl (Am), and northern saw-whet owl (Am) are considered rare or accidental and the flammulated owl (Us), snowy owl, and barred owl are listed as sensitive species. Some of the largest populations of burrowing owls in the state of Washington reside at the Hanford Site. Burrowing owls nest in five to ten feet long holes dug by badgers or other mammals in open sagebrush, old fields, and culverts in roadcut banks. Burrowing owls consume rodents, frogs, small birds, and insects as their main prey items. The family remains together until September when they migrate as far south as South America during the winter. Great horned owls (Ur), long-eared owls (Ur), barn owls (Ur), and short-eared owls (Ur) also nest at the Hanford Site. Several pairs of long-eared owls nest in the riparian zone near BC Reactor. Great horned owls and barn owls also reside in the old reactor facilities near the Columbia River.

(Top Clockwise) Great horned owl, Long-eared owl, Short-eared owl. (Bottom left) Young burrowing owls, (bottom right) Barn owl.

Strigiformes

OWLS

nighthawks

Common night hawk

Caprimulgiformes

This order is represented by the family CAPRIMULGIDAE and two species, the common nighthawk (Cs) and common poorwill (Us). Nighthawks and poorwills prey on flying insects and both nest at the Hanford Site. Nighthawks are widespread in their distribution, whereas the poorwills seem to be restricted to the upper elevations of Rattlesnake Mountain.

White throated swift in flight

Apodiformes

This order is represented by the *swifts* (APODIDAE) and *hummingbirds* (TROCHILIDAE) for a total of five species.

White-throated swifts (Rs) inhabit the cliff areas on Umtanum ridge, where they also nest. Four species of *hummingbirds* (black-chinned (Am), calliope (Um), rufous (Um), and Anna's (Am)) migrate through the Hanford site in the spring and fall. The calliope and rufous hummingbirds are the species most often observed.

hummingbirds & swifts

West Lake below Gable Mountain

COYOTE AND HUMMINGBIRD

Coyote was always wandering around, and now he went upriver. Suddenly he heard a voice saying, "Whoever is going upriver, let us murder one another."

Coyote said, "What?"

"Whoever is going upriver, let us murder one another," the voice said again.

Coyote said, "I wonder who that is." Then something hit him suddenly in the head. It was Hummingbird. It quickly killed Coyote and then threw his body in the river.

Coyote drifted downstream a bend and a half and then drifted ashore. There Magpie was flying around. "Oh, there's Coyote. What happened to the poor thing," he said. He landed on him and pecked his eyes. Coyote woke up.

"Ah ahaah ah… you disturbed me," Coyote said. "I was dreaming that I helped girls across the headwaters. I was helping a chief's daughter get into a canoe."

"Where and whom were you helping across?" said Magpie. "That Hummingbird upriver killed you. Now here is how you can get revenge. The Hummingbirds live up on the hill, and they have a white feather hanging high from a tree up there. That is their heart. When you go upriver, they will say twice, 'Whoever is going upriver, let's kill each other.' As soon as they say that the second time, you answer, 'Ho!' and run as fast as you can, speeding up to where the white thing shines. Run and get hold of that white feather, pull it down, and you will kill them."

The Coyote said, "That's what I thought I would do."

Magpie said, "You never did think anything like that. Now go on. Do it exactly right, or they will kill you."

So Coyote went upriver again, and heard the voice say, "Whoever you are that goes upriver, let's murder one another." The second time the voice said this, Coyote said, "Ho!" and ran up the hill to the feather.

"Oh, Coyote is back!" the Hummingbirds cried. Coyote quickly grabbed the feather and broke it. Then all the Hummingbirds fell on their backs, dead. Thus, Coyote conquered the feared killers.

kingfishers

Coraciformes

This order has one family (ALCEDINIDAE) and one species (belted kingfisher (UR)). Kingfishers nest along the Columbia River and will dive into the water to capture small fish.

woodpeckers

Piciformes

This order is represented by one family of woodpeckers (PICIDAE) which is represented by four species. The Lewis' woodpecker (RM), downy woodpecker (RW), and hairy woodpecker (RW) are rare or migrants. The Lewis' woodpecker is listed as a state candidate species and are very abundant near Toppenish, Washington and Satus Creek. The northern flicker (CR) is the most abundant species and nests in tree cavities along the Columbia River.

(Left) Northern flicker, (Top right) Lewis' woodpecker, (Bottom) Downy woodpecker

passerines

Sage sparrows prefer to nest in sagebrush habitat and the Hanford Site contains the largest populations of this species in Washington.

Passeriformes

This is the largest order of birds found at Hanford, represented by 18 families and 107 species.

TYRANNIDAE:
Tyrant Flycatchers

The tyrant flycatchers are represented by 13 species, seven of which are considered rare migrants (RM)—(olive-sided, willow, and dusky) or accidental migrants (AM)—(Hammond's, gray, Pacific slope, black phoebe). Ash-throated flycatchers (Rs) are generally observed near water in the summer. Species that are known to nest at the Hanford Site are the Say's phoebe (Us), western kingbird (Cs) and eastern kingbird (Us). Flycatchers are characterized by capturing insects in the air. The kingbirds are also not hesitant in chasing away larger birds such as raptors and ravens if perceived as a threat.

The other species which are uncommon migrants (UM) are the western wood-pewee and Cordilleran flycatcher.

(left) Western kingbird, Western kingbird nest, (right) Eastern kingbird.

Alaudidae: *Larks*

The only species in this family is the abundant horned lark (CR), which is a year round resident and constructs ground nests. They feed on insects and seeds, and in some years may rear two sets of young.

(Top left to right) Typical horned lark ground nest; Juvenile horned lark in ground nest (Bottom, clockwise) Cliff swallows gathering mud for nest building; Cliff swallow nests located on basalt cliff face; Northern rough-winged swallow. Typical bank swallow nests in side of steep bank

Hirundinidae: *Swallows*

The *swallows* are represented by six species, four of which nest at the Hanford Site (northern rough-winged (Us), bank (Us), cliff (Cs), and barn (CR). The cliff and barn swallow are the most common species which construct nests of mud. Many of these nests are constructed on the sides of buildings throughout the site. Bank swallows inhabit localized areas along the Columbia River where they may nest along steep banks. Tree swallows (Um) and violet-green swallows (Um) are occasionally observed.

passerines

Black-billed magpie

CORVIDAE:
Jays, Magpies, & Crows
This family is represented by six species of jays, magpies, and crows. The black-billed magpie (CR) and common raven (CR) are nesting species that are year-round residents and have reputations as scavengers and often feed on animal carcasses along roads. The four other species (Steller's jay (RW), scrub jay (AM), Clark's nutcracker (AM), and American crow (RR)) are considered rare or accidental. American crows can be observed at the Vernita Bridge rest stop area on the Columbia River and are also common in nearby towns.

BLUE JAY AND THE WELL-BEHAVED MAIDEN

Once there lived a maiden, a chief's daughter, who was very hard to woo. One day she told her father, "I'm not going to marry anyone foolishly. He will be the one who has big calves and nice legs. That's the one who will be my husband."

So the Chief said to everyone, "This is the plan my daughter has, whoever has good legs and big calves, he is the one who will marry." Then all the men who were there stuck their legs into the girl's teepee so that she could look at them. But she did not like any of them and told them to go away.

Now there was a young man, Bluejay, who had a plan on how to win the girl. He stuffed his legs with pine moss so that he looked as if he had big calves. Then he stuck his legs inside the girl's teepee. When she saw those legs, she looked at them for a long time. Finally, she said, "I have never seen legs like these! He has nice big calves! I will have you for a husband."

Then she gathered up her clothes and moccasins which she had been sewing with porcupine quills. She left the girl's lodge and followed the young man.

The two of them came to a stream. The girl took off her moccasins and began to wade across the stream. The boy didn't take his moccasins off, though. He just walked right in. As they were wading, the girl felt something get tangled around her legs. "Where is this pine moss coming from?" she asked. It was the moss that was coming out of his legs.

When they reached the other side, the girl saw that Bluejay's legs were nothing but bones. There were no fat calves on them. The girl felt terrible, but she could not go back for she had come this far with him. He was now her husband.

Bluejay and his new wife reached the tent. She saw that there was hardly any food there, only pine gum piled up. So that is all she had to eat.

One day Bluejay said, "Maybe the girl needs something other than pine gum to eat. Maybe the wild turkey I shot last summer will do."

His grandmother said, "I don't know how she has lived even this long. Why don't you give her the fat from the knee of an eagle and see if that satisfies her hunger!" Then the grandmother told him, "Let us try to do some wedding trading. At least we have pine gum to trade." So they filled their packs with gum and traded the best they could.

This well-behaved maiden made a big mistake. Good looks last only a short while. Then you may live in poverty. But if you do your best to overcome your mistake, you could be rewarded.

Stellar's jay

passerines

PARIDAE: *Titmice*
The black-capped chickadee (Uw) is the only member of this family. The highest populations of these birds at the Hanford Site occur in the winter.

SITTIDAE: *Nuthatches*
The red-breasted nuthatch (Rr) is the only member of this family and is rarely observed.

CERTHIIDAE: *Creepers*
The only member of this family documented at Hanford is the brown creeper (Am) which is considered accidental.

(Top) Marsh wren, (Bottom) Canyon wren.

THE WREN

The Earth people wanted to make war on the Sky people. Grizzly Bear was the chief of the Earth people, and he called all the warriors together. They were told to shoot in turn at the moon (or sky). Everyone's's arrows fell short. Only Wren had not shot his arrow. Coyote said "He doesn't have to shoot because he is too small and his bow and arrows are too weak." But Grizzly Bear said that Wren must shoot. Wren shot his arrow and it hit the moon and stuck fast. Then the others shot their arrows, each of which stuck in the neck of the preceding one until they had made a chain reaching from the ground to the sky. Then everyone climbed up, Grizzly Bear going last. He was heavy and his weight broke the chain, but he made a spring and caught the part of the chain above him and this caused the arrows to pull out at the top where the leading warriors had made a hole to enter the sky. The chain fell down and left the people without means to come down. The Earth people attacked the Sky people and defeated them in the first battle but the Sky people in the next battle were victorious. The defeated Earth people ran for the ladder but many were killed on the way. When they found the ladder broken each prepared himself the best way he could so as not to fall too heavily and one after the other they jumped down. Flying Squirrel was wearing a small robe, which he, which he spread out like wings when he jumped. Therefore, he has something like wings now. Whitefish looked down the hole before jumping, but puckered up his mouth and drew back when he saw the great depth. Therefore, he has a small puckered mouth to the present day. Sucker jumped down without first preparing himself and his bones were broken, and that is why sucker's bones are found in all parts of its flesh now.

TROGLOGYTIDAE: *Wrens*
This family is represented by six species of wrens, most of which are considered uncommon or rare. Rock wrens (Us) and canyon wrens (Rs) are found on Gable Mountain, Rattlesnake Mountain, and Umtanum ridge on the basalt cliffs and outcrops. Marsh wrens (Ur), house wrens (Rs), Bewick's wrens (Rs), and winter wrens (Am) prefer brushier areas, riparian zones, and marshes.

MUSCICAPIDAE: *Kinglets, bluebirds, & thrushes*
This family of nine species is represented by the kinglets, bluebirds and thrushes. Only the American robin (Cr), which nests at the site, is common. The western bluebird (Rm), mountain bluebird (Rm), Townsend's solitaire (Rm), golden-crowned kinglet (Uw), ruby-crowned kinglet (Uw), Swainson's thrush (Rm), hermit thrush (Uw), and varied thrush (Rw) are all uncommon or rare. The western bluebird is a state monitor species.

(Clockwise) Rock wren, American robin, Mountain bluebird. (Below) Wildflowers on Umtanum Ridge.

passerines

53

MIMIDAE: *Mockingbirds & thrashers*

This family is represented by the gray catbird (Am), northern mockingbird (Am) and sage thrasher (Us). The sage thrasher, which is a state candidate species, has nested on the Fitzner–Eberhardt Arid Lands Ecology Reserve (ALE) and the Wahluke Slope. Northern mockingbird observations are becoming more common, especially on the North Slope.

(Top) Sage thrasher (Middle left to right) Bohemian waxwing, Loggerhead shrike (Bottom) Prickly pear cactus.

BOMBYCILLIDAE: *Waxwings*

The Bohemian waxwing (Rw) and cedar waxwing (Cs) are the two species represented in this family. Both species are normally observed in the winter, with the cedar waxwing being more common. Cedar waxwings are also observed in the springtime.

MOTACILLIDAE: *Pipits*

This family is only represented by the American pipit (Cm), which is a common migrant.

LANIIDAE: *Shrikes*

The northern shrike (Rw) and loggerhead shrike (Cs) make up this family. Northern shrikes are rarely observed in the winter. Loggerhead shrikes are listed as a federal candidate species. Loggerhead shrikes are predatory birds which nest in shrub-steppe habitat building bulky nests of sticks and grasses five to twenty feet above the ground. Shrikes kill mice by breaking their necks with a few deft strokes of the bill. They are also proficient at catching insects and lizards. They usually impale the catch on a thorn, cactus, or a barbed wire fence.

STURNIDAE: *Starlings*

The European starling (Cr), which is an introduced species and distributed nationwide, is the only species in this family.

VIREONIDAE: *Vireos*
This family is represented by five species of *vireos* (solitary (Um), Hutton's (Am), warbling (Um), Philadelphia (Am), and red-eyed (Um)) all of which are accidental or uncommon migrants. The warbling vireo and solitary vireo are the two species most often observed in riparian zones during spring and fall migrations.

EMBERIZIDAE:
Emberizids
This large family is made up of the wood warblers, tanagers, grosbeaks, buntings, towhees, sparrows, longspurs, blackbirds, and orioles.

There are 13 species of *warblers*, 12 of which are considered uncommon: (orange crowned (Um), yellow (Us), Townsend's (Um), yellow breasted chat (Us), MacGillivray's (Um), Wilson's (Um)); rare: (Nashville (Rm), common yellowthroat (Rm)); or accidental migrants (Am): (blackpoll, Tennessee, palm, American redstart). The species most often observed at Hanford are the yellow warbler and yellow-rumped or Audubon's warbler (Cw).

The western tanager (Um), rose-breasted grosbeak (Am), black-headed grosbeak (Us), lazuli bunting (Us) and rufous-sided towhee (Uw) are considered uncommon, rare or accidental. Black headed grosbeaks and lazuli buntings have been known to nest in riparian areas along the Columbia River and on the North slope.

(Right) Lazuli bunting (below) Rufous-sided towhee (Bottom left) Brewer's blackbird.

Fifteen species of sparrows have been observed at the site. *Sparrows* which nest at the Hanford site are the Brewer's (Rs), vesper (Rs), lark (Us), sage (Cs), savannah (Us), grasshopper (Rs), and song (Cr). Other species observed at the site include the black-throated (Am), chipping (Rm), fox (Rm), Lincoln's (Rm), swamp (Am), and Harris' (Rw). There are indications that black-throated sparrows may be becoming more common at the site as observations seem to be increasing.

The sage sparrow and grasshopper sparrow are state

passerines

monitor and candidate species. The largest populations of sage sparrows in the state of Washington can be found at the Hanford Site. The sage sparrow only lives and nests in sagebrush and as such is extremely vulnerable to loss of sagebrush habitat. Sage sparrows arrive at Hanford in late April and early May and leave in late August. They live on the ground under the desert shrubs. Sage sparrows do not scratch for food, but run around looking for grasshoppers, bugs, beetles, ants, spiders or seeds. Nests are most often built in sagebrush, a few inches to three and one-half feet above ground. Grasshopper sparrows are not abundant at Hanford but they do nest on the ALE site.

(Top clockwise) White crowned sparrow, Sage sparrow, Black-throated sparrow. (Bottom) Lark sparrow, Chipping sparrow.

Other common members of this family which nest at wetland and riparian sites include the red-winged blackbird (Cr), western meadowlark (Cr), yellow-headed blackbird (Us), Brewer's blackbird (Ur), brown-headed cowbird (Ur), and northern oriole (Us). Species that are not commonly observed include the laplund longspur (Am), snow bunting (Am), bobolink (Am), and rusty blackbird (Am).

Fringillidae: Finches & Siskins

This family is represented by the rosy finch (Rw), purple finch (Am), house finch (Cr), Cassin's finch (Um), American goldfinch (Ur), lesser goldfinch (Am), pine siskin (Rw), common redpoll (Aw), and evening grosbeak (Rw). The house finch and the American goldfinch, which nest at the site, are the most common species.

Passeridae: Old World sparrows

The introduced house sparrow (Cr), which nests at the site, is the only species in this family.

(Top) Western meadowlark, (Bottom clockwise) Red-winged blackbird, Brown headed cowbird, Yellow-headed blackbird, Evening grosbeak

MAMMALS

The Trees Now Shed Their Leaves, And The Days Grow Colder

Nez Perce Thanksgiving Celebration

shrews

Insectivora

This order consists of the family SORICIDAE and two species of *shrews*, the vagrant (UR) and the Merriam's (UR). Both species prey on insects. The Merriam's shrew is found at higher elevations on Rattlesnake Mountain and the vagrant shrew seems to prefer lower elevations in riparian zones. The Merriam's shrew is a state candidate species. It is not uncommon to find Merriam's shrews in the diets of long-eared owls and burrowing owls.

Chiroptera

(Clockwise) Silver-haired bat, Little brown bat, Pallid bat. (Bottom right) Little brown myotis

bats

This order is represented by six species of bats in the family VESPERTILIONIDAE, all of which are considered federal candidate species. Bats have not been studied in much detail at Hanford. Battelle scientists have speculated that up to 14 species of bats could potentially occur at Hanford. Several species have been captured in the Gable Mountain, Snively Canyon, and Rattlesnake Springs areas. A colony of more than 100 pallid bats was found in the 100-F Reactor Building. Other species of bats including small-footed myotis, silver-haired bats, Yuma myotis, little brown myotis, and western pipistrel have also been collected in the reactor buildings along the river. Bat surveys conducted by the Washington Department of Fish & Wildlife have indicated that several of the bat species mentioned above and others have been observed in counties north of the Hanford Site.

carnivores

(Top) Cougar, (Bottom) Raccoon

Carnivora

The carnivores are represented by four families which include the raccoons (PROCYONIDAE), weasels, skunks, badgers, otters (MUSTELIDAE), coyotes (CANIDAE), and bobcats and cougars (FELIDAE).

PROCYONIDAE:
Raccoons

The raccoon (RR) is the only member of this family. It prefers riparian areas where it forages for food. Raccoons are considered uncommon at the Hanford Site but in general are becoming more abundant throughout the Columbia Basin and especially along the Yakima River.

RACCOON-BOY AND HIS GRANDMOTHER

There lived a Raccoon-Boy with his grandmother. One day, the old woman brought some roots called khouse. The next evening, Raccoon-Boy went next door and broke every root and ate all the food from inside every one of them. Then he put the root shells back together again, having put sand and dirt inside.

A little while later, Grandmother got hungry. She went to eat some of the roots, but they were all stuffed with dirt. "He is always doing this," Grandmother said. "He always leaves me out." When the boy came home, she picked up a piece of burned wood and paddled him.

Raccoon-Boy went crying down the hill. After a while, he started looking for crawfish. But then he sensed that someone was there. Suddenly Grizzly Bear was standing right behind him. Raccoon-Boy ran as fast as he could to the top of a tree, but Grizzly Bear climbed up after him. She had with her a big sack with needles stuck inside it. She planned to put Raccoon-Boy inside there and kill him. "What are you running away from, my nephew?" Grizzly said to Raccoon-Boy. "Why don't you look for lice in my hair?" So there up high, Raccoon-Boy began to look for lice. Grizzly had frogs for lice. Raccoon-Boy made a sound, "taq!" (SLAP) with his hand. "What are you doing?" Grizzly said. "You are doing it the wrong way. Hand me one and I'll show you." So Raccoon-Boy handed her one. She made a sound "kikh-kikh" (SQUISH! POP!). "Do it this way," she told Raccoon-Boy. "Do it with your teeth." It turned Raccoon-Boy's stomach.

Then Raccoon-Boy caught one and chewed on it and made the same sound, in order to please Grizzly Bear. She said "That's right! Now you're doing it right!"

Raccoon-Boy caught another one, then touched Grizzly's ear, and said, "Oh, one of the lice ran into your ear."

"Get a needle out of the pack and take it out," Grizzly said.

"Gladly," answered Raccoon-Boy. He took the needle and pretended to look for the louse, but then he stuck the needle in her ear and pushed hard. She fell down and down and hit the ground with a thud. Then she lay there shaking all over.

Raccoon-Boy stayed up in the tree, afraid to come down. He threw things down on top of Grizzly to see if she was dead. Finally he thought, "Now she is dead," and climbed down out of the tree.

Then he ran to his grandmother's house. "Grandmother, I have killed Grizzly Bear! he said.

"Now how could you kill anything as mean as Grizzly Bear?" his grandmother said.

"Come and see," he replied. And they went and saw it lying there. They brought it back and cut it up and roasted it. Then the old woman accidentally cut her finger. Raccoon-Boy said, "Now you are menstruating. Make a teepee, away from the people. You might bring bad luck to my kill."

She said, "What? I just cut myself."

"No, no, go to the menstrual lodge. Hurry," Raccoon-Boy insisted. So she did.

When the meat was done, Raccoon-Boy took it out. Then he began to eat. Then, in a loud voice he began talking, pretending that people had just arrived for a visit.

"I wonder who came over to visit?" Grandmother said from the menstrual teepee.

Raccoon-Boy was saying loudly, "Yes, take some meat! Here, have some more! Take the rest home with you!" Then he pretended that they were leaving. "Oh, you forgot your dog! Good-bye!" Then he mimicked, "Bow Wow. Bow wow."

When the grandmother came back in the house, Raccoon-Boy said, "Visitors were here, Grandmother. They took all the meat."

The old woman was hungry, and walked around looking for some meat. She found some marrow on the ground and picked it up. Then the boy said, "Let me see that." When she gave it to him, he sucked it all out.

Then the woman felt sorry for herself. "That's the way he has been treating me. Again he is leaving me out. There really weren't any people here. I will leave him." She didn't say anything to him, but picked up the bearskin she had, and left.

Raccoon-Boy was all filled up, and now he started craving for sweets. But his grandmother didn't show up to make him anything like she always did. When he last saw her, she was walking up the hill with a root digger and a grizzly bear hide.

Badger

MUSTELIDAE:

Weasels, badgers, & skunks
This family is represented by the mink (Rr), otter (Rr), badger (Cr), long-tailed weasel (Ur), short-tailed weasel (Aw), and striped skunk (Rr). The badger is the most abundant member of this family and its main prey items are Townsend ground squirrels and Great Basin pocket mice. Badgers capture small mammals by digging with their powerful front claws. Abandoned badger dens may be utilized by burrowing owls. Long-tailed weasels are usually observed in riparian zones, where they prey on small mammals. Skunks, mink, otters, and short-tailed weasels have only been observed a few times.

(Clockwise) Long-tailed weasel, badger, badger den, otter, striped skunk, short tailed-weasel

carnivores

carnivores

(Top) Coyote, (Bottom) Bobcat

CANIDAE:
Coyote
This family is represented by the coyote (CR), the most abundant carnivore at the Hanford Site. Coyotes are opportunists and prey upon small mammals, Canada geese, rabbits, and carp. Coyotes also prey upon mule deer fawns in the spring and will even swim out to the islands in the Columbia River where many of the deer are born.

FELIDAE: *Cats*
This family is represented by the bobcats (UR) and cougars (AM). Bobcats are seldom observed but they are known to reside on Rattlesnake Mountain, Gable Mountain, and at some of the abandoned reactors along the Columbia River. In 1993 a Hanford worker reported sighting a cougar that had crossed the road near Gable Mountain. Biologists who subsequently followed up on this sighting reported cougar tracks in the soft sand. It is not known at this time whether or not cougars reside at the site or merely travel through the area. A cougar was also observed by two policeman north of Richland in 1996.

Coyote Finishes his Work

From the very beginning, Coyote was traveling around all over the earth. He did many wonderful things when he went along. He killed the monsters and the evil spirits that preyed on the people. He made the Indians, and put them out in tribes all over the world because Old Man Above wanted the earth to be inhabited all over, not just in one or two places.

He gave all the people different names and taught them different languages. This is why Indians live all over the country now and speak in different ways.

He taught the people how to eat and how to hunt the buffalo and catch eagle. He taught them what roots to eat and how to make a good lodge and what to wear. He taught them how to dance. Sometimes he made mistakes, and even though he was wise and powerful, he did many foolish things. But that was his way.

Coyote liked to play tricks. He thought about himself all the time, and told everyone he was a great warrior, but he was not. Sometimes he would go too far with some trick and get someone killed. Other times, he would have a trick played on himself by someone else. He got killed this way so many times that Fox and the birds got tired of bringing him back to life. Another way he got in trouble was trying to do what someone else did. This is how he came to be called Imitator.

Coyote was ugly too. The girls did not like him. But he was smart. He could change himself around and trick the women. Coyote got the girls when he wanted.

One time, Coyote had done everything he could think of and was traveling from one place to another place, looking for other things that needed to be done. Old Man saw him going along and said to himself, "Coyote has now done almost everything he is capable of doing. His work is almost done. It is time to bring him back to the place where he started."

So Great Spirit came down and traveled in the shape of an old man. He met Coyote. Coyote said, "I am Coyote. Who are you?"

Old Man said, "I am Chief of the earth. It was I who sent you to set the work right."

"No," Coyote said, "you never sent me. I don't know you. If you are the Chief, take that lake over there and move it to the side of that mountain."

"No. If you are Coyote, let me see you do it."

Coyote did it.

"Now move it back."

Coyote tried, but he could not do it. He thought this was strange. He tried again, but he could not do it.

Chief moved the lake back.

Coyote said, "Now I know you are the Chief."

Old Man said, "Your work is finished, Coyote. You have traveled far and done much good. Now you will go to where I have prepared a home for you."

The Coyote disappeared. Now no one knows where he is anymore.

Old Man got ready to leave, too. He said to the Indians, "I will send messages to the earth by spirits of the people who reach me but whose time to die has not yet come. They will carry messages to you from time to time. When their spirits come back into their bodies, they will revive and tell you their experiences.

"Coyote and myself, we will not be seen again until Earth-woman is very old. Then we shall return to earth, for it will require a change by that time. Coyote will come along first, and when you see him you will know I am coming. When I come along, all the spirits of the dead will be with me. There will be no more Other Side Camp. All the people will live together. Earthmother will go back to her first shape and live as a mother among her children. Then things will be made right."

Now they are waiting for Coyote.

Rodentia

rodents

(Top) beaver, (bottom left) Yellow-bellied marmot, (bottom right) Townsend Ground Squirrel

The rodents are represented by seven families and 16 species.

S<small>CIURIDAE</small>:
Ground squirrels
This family is represented by the Townsend ground squirrel (U<small>R</small>), Yellow-bellied marmot (R<small>R</small>), and least chipmunk (A<small>M</small>). Townsend ground squirrels have very localized distributions and on the 200 Area Plateau it is not common for small colonies to persist only for a short time. These squirrels are more common on the A<small>LE</small> Reserve and 300 Areas and are prey to badgers and hawks. They prefer to eat green vegetation like balsam root and they hibernate during the hot spring and summer months. Yellow-bellied marmots are found only on Rattlesnake Mountain and Umtanum Ridge. The least chipmunk is rare and has only been observed a few times on the A<small>LE</small> Reserve.

(Left) Northern pocket gopher, (right) Great Basin pocket mouse.

GEOMYIDAE:
Pocket gophers

The northern pocket gopher (UR) is the only member of this family. The highest populations of pocket gophers are found on Rattlesnake Mountain. They are also found on Gable Mountain and in the Columbia River riparian zone. They are uncommon on the Central plateau.

HETEROMYIDAE:
Pocket mice

The Great Basin pocket mouse (CR) is the only member of this family. This species is the most abundant small mammal found at the Hanford Site. It lives almost exclusively on seeds and is a prey item for hawks, owls, coyotes, badgers, and snakes. Pocket mice construct distinctive mounds which are plugged in the daylight hours. Pocket mice are active at night when they come out to search for seeds.

(Left) Pocket gopher cast, (below) beaver.

CASTORIDAE:
Beavers

The beaver (RR) is the only member of this family and the populations along the Columbia River are not high. They are more common from the Hanford townsite downstream to Richland.

(Clockwise) Northern grasshopper mouse, Muskrat, Sagebrush vole, Pocket mouse mound, Bushy-tailed woodrat.

<u>CRICETIDAE</u>:
woodrats, voles, & muskrats
This family is made up of seven species. The deer mouse (CR) is common throughout the Hanford Site. The western harvest mouse (UR) prefers riparian areas, and the northern grasshopper mouse (RR) and sagebrush vole (RR), which are state monitor species are more common on the ALE Reserve and seem to be closely associated with sagebrush habitat. The montane meadow vole (RR) is only found near water and the bushy tailed woodrat (UR) is found in basalt outcrops and debris piles next to the Columbia River. Muskrats (CR) are common in the Columbia River.

<u>MURIDAE</u>:
house mice & rats
This family is represented by the house mouse (CR) and Norway rat (UR). These species are considered pest species and reside in and around the many buildings at the Hanford Site.

rodents

ERETHIZONTIDAE:
porcupines

This family is represented by the porcupine (UR). Porcupines are most often observed at the old Hanford townsite and along the Columbia River where there are enough trees to provide food and cover. Their main food item is tree and willow bark. They are the occasional prey of cougars, bobcats, and coyotes.

Porcupine

SNOWSHOE AND COTTONTAIL RABBIT

Once upon a time, Snowshoe Rabbit and Cottontail Rabbit, who were close friends, or "brothers" were living together. Everyday they would go roaming around, this way and that. One day, Snowshoe Rabbit was up in the mountains, when it suddenly began to snow. The snow got so deep, finally, that he couldn't get back home. So he had to spend the winter in the mountains, but Cottontail Rabbit stayed in the valley for the winter.

The next spring, when they met again, Snowshoe Rabbit said, "Well, my friend, you have gone through a hard winter. When I would look out this way toward the valley, it would be dark over there and look as if it were raining. I used to tell myself, 'I wonder how my friend is passing his time, and where he is.'"

Cottontail Rabbit said, "That's the same thing I would do. I would look toward the mountains and watch. It was dark with storms as if the rain and snow were pouring down. I wondered how you were living."

"My friend, you thought wrong," Snowshoe Rabbit said. "I had a good home, and would throw good wood into the fire to burn. I would lie with my back toward the fire, as it turned into charcoal to make the house warm and comfortable. I gathered lots of food and had plenty to eat. My living was very pleasant in the mountains, but I worried about you."

Cottontail Rabbit said, "Well, friend, you worried about me for another. As you see from here, I have a good house where there are plenty of loose rocks. I would throw chokecherry wood into the fire, and it would burn to charcoal. Then I would warm my back toward it. I, too, lived well."

Snowshoe Rabbit said, "Yes, I see how it was for you. Let us agree, then, that you, Cottontail, will live in the lower country, and I will live in the mountains. We have learned that my best living is in the mountains and your best living is in the low country. At the beginning of each spring and winter season, I will change my clothing. When it snows, I will dress in the same color as the white snow, so no one can see me or find me. But when spring comes, I shall put on new gray clothing so that nothing can find me easily."

Snowshoe Rabbit has never come to the low country since then. On the other hand, Cottontail Rabbit is always found in the low country. That is how they live.

Lagomorpha

rabbits

This order is represented by 3 species in the family LEPORIDAE. The Nuttall's cottontail (CR) is found around buildings and some of the Columbia River riparian zones throughout the Hanford Site. The white-tailed jackrabbit (RR) is observed infrequently on Rattlesnake Mountain. The black-tailed jackrabbit (CR) is common in sagebrush habitat throughout the site. All of these species are prey for hawks, owls, bobcats, badgers, and coyotes.

*White-tailed Jackrabbit.
Facing page: Cottontail Rabbit*

COTTONTAIL BOY AND THUNDER

Once upon a time Cottontail Boy and his grandmother were living together. Cottontail Boy was somewhat disobedient one day, and his grandmother told him to behave.

"Stop it, grandchild. If you don't behave, Salmon Pemmican Grandchild will come and say, 'Here is Salmon Pemmican Grandchild.' Then he'll grab you and take you away," his grandmother warned.

Just at that time Grizzly Bear was wandering around nearby. Grizzly Bear wondered why they were scaring each other. Just as Cottontail Boy was getting pretty mad, Grizzly Bear reached inside the house saying, "Here is Salmon Pemmican Grandchild."

Cottontail Boy grabbed Grizzly Bear's arm and tore it off, taking it away. Grizzly Bear reeled round and round and said to Cottontail Boy, "Boy, prop my arm up," and finally he dropped over dead.

Then they baked Grizzly Bear, and Cottontail Boy told his grandmother to tan his hide. He skinned the arm and his grandmother tanned it for him. Then he decorated it after it dried.

Soon he said to his grandmother, "Grandmother, now I'm going to go steal Thunder's Wife."

She said to him, "Calm down, boy. Don't say such a thing. It scares me. How could you ever take his wife away?"

"I don't care what you say, I'm going," said Cottontail Boy.

So he went on up to where Thunder's wives were, and he found them. They were digging for roots. He asked one of them, "Who is the favorite wife?"

She pointed out the favorite to him. That one had one-half of her dress painted red and the other had painted yellow. Cottontail Boy went up to her and said, "Throw away your digger."

The woman laughed saying, "How funny! He tells me to throw my digging stick away."

But her friends told her to throw away the stick and see what he would do. And so she did.

Then Cottontail Boy said, "Throw away your digging bag."

She did that too.

"Throw away the child." She had a child on her back. "Throw away the child and put it down here."

Then he told her, "Now we are going away together."

"Don't say such a thing!" the woman cried.

Then he grabbed her by the arm and dragged her, running along. She tried to escape, but to no avail. Finally, she gave up and she said that she would go with him.

Meanwhile the other women hurried to bring the news to Thunder.

"Listen! Your wife has been stolen by Cottontail Boy!"

Thunder got ready and rushed to save his wife.

When Cottontail Boy saw Thunder coming, he hid the woman by lying on his back on top of her.

"Be still!" he warned her. Just then Thunder sent hail to fall so that Cottontail Boy would blink, and Thunder would have a chance to strike him. But Cottontail Boy glared angrily at Thunder. Even though Cottontail Boy tried in various ways, nothing happened. Finally he pulled out Grizzly Bear's arm, all decorated. He aimed it at Thunder and hit him with it.

"I'm just as dangerous as you!" cried Cottontail Boy.

He went on knocking Thunder, until finally Thunder said, "You've defeated me. You are truly dangerous. You have won my wife. Take her. Now I am going back."

So Cottontail Boy took his bride back to his grandmother.

It was in this way that the name of the place, Pipaxliwam, originated. Its formation shows the way Thunder hit it that day. And, it is there to this day; there is ice there inside a cave.

Hunting blinds such as the one to the left were used historically by Native Americans to hunt large game such as deer, elk, and antelope. The blinds were normally constructed on talus slopes adjacent to game trails and animal migration routes. Hunters would often lay in these blinds for several hours to several days waiting for prey. Many of these hunting sites still exist throughout the Columbia Basin region.

elk & deer

(Clockwise) Mule deer bucks, Elk herd, Mule deer doe and fawn.

Artiodactyla

CERVIDAE

This family is represented by the mule deer (CR), white-tailed deer (RR), and elk (CR). Mule deer are common at the Hanford Site and have a widespread distribution. Some of the largest populations are found in the 100 Areas proximal to the Columbia River. White-tailed deer are rare or accidental. There is a large herd of about 300 elk that reside on the ALE Reserve. This herd started in 1972 when a few animals presumably showed up from the mountains to the west. They are commonly observed in Snively canyon and Rattlesnake Springs where there is running water. Many of the elk have been harvested by hunters when they stray over the Hanford Site boundaries.

Mule deer fawn

> The white men were many and we could not hold our own with them. We were like deer. They were like grizzly bears. We had a small country. Their country was large. We were contented to let things remain as the Great Spirit made them. They were not, and would change the rivers if they did not suit them.
>
> *Chief Joseph, Nez Perce*

Coyote and White-tailed Buck

When the different kinds of deer were created, there was White-tailed Buck. Coyote used to see him sitting there. Nothing disturbed him, even when Coyote came over and tried to scare him by shouting in various ways. He was just peaceful and sat chewing his cud. For a long time Coyote studied the matter, wondering, "How can he become more alert? He is too indifferent. Anyone, even a woman, could club him to death." In this way Coyote contemplated now.

Then he thought, "Maybe this will do it," and pointed his genitals at White-tailed Buck's nose, almost touching it. Buck got the scent and gave a warning snort. After that, whenever Coyote showed up, Buck snorted. "There, you reacted in the right way," Coyote said. "That's what will make you wary. Only a man who prepares himself, taking a sweat-bath and cleansing himself, will be able to kill you. But not those who do not bathe. You were just too complacent, so much so that even women could kill you. That's the way you were. But this is the way you will be from now on." From that time on, white-tailed bucks became difficult to approach. Only those who are prepared properly have a chance to kill them.

Elk

REPTILES & A

MPHIBIANS

> There is an awful lot of power in the rocks.
> We call them the old men because those rocks
> have been here a long, long time.
> *Gilbert Towner*

reptiles

Reptiles have not received much study at the Hanford Site and are not often observed. Reptiles are represented by 9 species that are classified as lizards, skinks, or snakes.

Lizards are represented by the sagebrush lizard (UR), side-blotched lizard (CR) and short-horned lizard (UR). Snakes that inhabit the Hanford Site include the gopher or bull snake (CR), rattlesnake (UR), desert night snake (RR), striped whipsnake (RR) and the green racer (UR). Skinks are represented by the Western skink (RR).

The side-blotched lizard is the most common and abundant lizard at the Hanford Site. It is widespread but seems to prefer sandy areas at the lower elevations on site. The sagebrush lizard is less common and seem to be associated with bitterbrush/Sandberg's bluegrass habitats.

The short-horned lizard can be found site wide but they seem to be more common at the upper elevations of Rattlesnake Mountain. They have also been observed several times on the stabilized sand dunes that exist on both sides of the Columbia river. All of the lizards are prey to snakes and birds such as the loggerhead shrike and common raven.

Rattlesnakes are distributed throughout the site but are most common on Rattlesnake and Gable Mountains, Gable Butte and areas where there are basalt outcrops. They have also been observed at several of the deactivated reactor sites that are near the Columbia River. The desert night snake is not observed very often but is most often seen around basalt outcrops. The striped whipsnake is the rarest snake at the site and has only been observed on very few occasions. The gopher or bull snake is the most common snake at the site and is a common prey item of ravens, Swainson's hawks, ferruginous hawks, red-tailed hawks, coyotes and badgers.

Green racers are not uncommon and can move very fast. They are also a prey item for hawks and mammalian carnivores.

Western skinks are very rare at the Hanford Site. A western skink was collected for radionuclide analysis at Hanford in 1952 but has never been included on any Hanford wildlife lists. Western skinks have been collected and photographed in the Columbia Basin by the United States Fish and Wildlife Service, so it would not be surprising to find them at the Hanford Site.

(Pg 76, top to bottom) Gopher snake, Short-horned lizard. (Pg 77 top, clockwise) Gopher snake, Desert night snake, Green racer, Rattlesnake. (Bottom, clockwise) Side-blotched lizard, Sagebrush lizard, Juvenile western skink.

amphibians

Amphibians have not received much study at the Hanford Site, as well. Like the reptiles, the amphibians are not often observed. Amphibians are represented by 6 species that are classified as frogs, toads, and turtles.

Frogs and toads are represented by the Pacific tree frog (RR), bullfrog (RR), western toad (RR), Woodhouse's toad (UR) and the Great Basin spadefoot toad (UR). Turtles are represented by the painted turtle (RR). The Woodhouse's toad is listed as a sensitive species.

Frogs and toads are not very common at the site and the western toad and bullfrog are very rare. Bullfrogs have been heard at some of the sloughs along the Columbia River. One western toad was collected near the H reactor during the summer of 1996. Pacific tree frogs are not very common but have been observed in the riparian zone of the Columbia River and elsewhere in the Columbia Basin. The Woodhouse's toad can also be found in the riparian zone of the Columbia River and has been observed several times between BC and D reactor. The Great Basin spadefoot toad stays buried underground and will come out to breed from May-July if there is enough available water. It can bury itself in sandy areas several miles away from the river. One species of frog that has been collected in the Columbia Basin is the northern leopard frog which has yet to be reported from the Hanford Site.

Painted turtles are uncommon at the Hanford Site, but are common in many of the Columbia Basin lakes and have been observed in some of the Columbia Basin sloughs.

(Pg 78, top to bottom) Painted turtle, Northern leopard frog. (Pg 79, clockwise) Bullfrog, Great Basin spadefoot toad, Pacific tree frog, Woodhouse's toad.

FROG AND BLUEJAY

Frog had a smooth pole set in the ground, and with it she had devised a means of killing off all the birds. A race would be run up the pole, and whoever got to the top first would cut off the other one's head. The pole leaned a little; and Frog would get on the upper side and make the opponent get on the underside. Thus, Frog continued to win races for a long time and managed to kill off many birds.

Coyote was in the camp. He became afraid that Frog was going to kill off all the birds. So Coyote gave a big feast and invited everyone to attend. He wanted the people to work out a plan to get the best of Frog, but everyone was afraid to run a race against Frog. After a scheme had been devised, Bluejay undertook the job; and Coyote made a big speech, calling everybody to the pole and announcing that there was to be a race between Frog and Bluejay.

Little Frog became uneasy and feared that Bluejay was going to win and then kill her. So when they were halfway to the top, Frog tried to kick Bluejay off the pole. When they were nearly to the top, Bluejay used his wings and flew the rest of the way. He got to the top first. When Frog got to the top, Bluejay kicked her, and she fell to the ground and was killed. Ever since that race there have been no feathers on the side of Bluejay's face because Frog had torn them all off when she tried to knock Bluejay from the pole.

After the race was over, Coyote made a speech, saying, "Hereafter there will be frogs on earth, but they will never hurt anyone. People will hear the frogs singing, and then they will know that warm weather is coming."

dangerous creatures

There are several animal species at the Hanford Site that are venomous or that can create other human health concerns. These include spiders, scorpions, bees, wasps, rattlesnakes, and rodents.

Hobo Spider

The hobo spider or aggressive house spider was introduced into the northwestern United States from Europe and it's bite is often confused as that of a brown recluse. Hobo spiders are also known as "funnel-web weaving spiders." Hobo spiders are common and widespread in the Pacific Northwest, but bites are relatively rare. These spiders are brown with gray markings and the legs and body are conspicuously hairy. The bite of the hobo spider is usually not felt and in most cases, the biting spider is not seen. Hobo spider poisoning initially produces an area of redness around the bite which often blisters, scabs, and takes several weeks to heal. Bites can produce serious systemic poisoning which includes severe headache, muscular weakness, visual disturbances, and hallucinations. First aid measures are ineffective in preventing the effects of Hobo Spider poisoning. Bites should be allowed to heal naturally and should be left exposed to the air whenever possible. Measures such as treating a bite with hot and cold packs and taking aspirin are not recommended. Adult Hobo Spiders are largely nocturnal, emerging in late July and August when the males leave their funnel webs in search of mates, sometimes entering houses in the process. Most bites occur from July through September.

Brown Recluse

The brown recluse is a medium sized spider up to 1/2 inch long and ranging in color from light fawn to dark brown. They have long, hair-covered legs that appear bare to the naked eye. The most distinguishing characteristic is a dark fiddleback shaped mark behind the eyes. Brown recluses do not reside in the Pacific Northwest nor at the Hanford Site but they have been collected when people travel from the Southwest to the Northwest. Since this spider is nonaggressive, most bites occur when someone accidentally brushes against one. Reaction from the bite ranges from mild to severe. The infected tissue usually dies and sloughs away leaving a sunken, ulcerating sore. The best treatment for the bite is immediate surgical removal of the affected area. An antidote is also available that sometimes prevents tissue destruction if administered within 48 hours.

Rattlesnake

Rattlesnakes are not uncommon in the appropriate habitat and there are many locations at Hanford where they may be found. Rattlesnakes may be found in the basalt outcrop areas on Umtanum Ridge, the ALE Site, and Gable Mountain and Gable Butte. A bite from a rattlesnake can result in serious poisoning with both local and systemic effects. Bites are seldom fatal and the best treatment is to get medical attention. Ice packs and tourniquets are not advised. Snake bites should be washed with soapy water.

(Pg 82) Brown recluse. (Pg 83) Barrel cactus, Rattlesnake.

Cottontail Boy and Rattlesnake

Once there lived Cottontail Boy and a small Rattlesnake Boy. They played all kinds of games together, and once while they were playing Rattlesnake thought to himself, "I wish I could swallow him. I wish he would quarrel with me."

But Cottontail Boy didn't quarrel, and they just went on playing. Then for some reason they finally got in a quarrel. Then Rattlesnake said, "You'll never escape—now I'm going to swallow you." And Cottontail Boy cried, "There's no reason why you should quarrel with me," and he wheeled around and ran up the hillside. Rattlesnake chased him up the hillside, his mouth hanging open and almost touching Cottontail's tail. Someone standing opposite the hill saw this, and suddenly, from across the river, cried, "Oh! They are chasing one another!" And he told them, "You will become a brown rock, and people will see you and say, 'There Rattlesnake is chasing Rabbit up the hillside.'"

dangerous creatures

Scorpion

Scorpions can be found throughout the Hanford Site. They prefer to come out at night, living underneath rocks and other debris during the day. The species of scorpion residing at the Hanford Site is not lethal to man. The sting of these scorpions is similar to that of a bee or wasp and should be treated the same way.

Tick

Ticks are found at the Hanford Site and seem to be more prevalent early in the spring. They are commonly found on sagebrush and Russian olives. Ticks are known to be carriers of Rocky Mountain Spotted Fever and Lyme Disease. Ticks should be removed by using blunt tweezers or covered fingers. Grasp the tick close to the skin and pull upward with steady, even pressure. Check to see that the entire tick has been removed and clean the area with antiseptic. If symptoms like headaches, fever, and chills develop, call a doctor immediately.

Bees & Wasps

Many species of bees and wasps reside at the Hanford Site. If stung the stinger should be removed by scraping a card across the wound (do not squeeze) and washed with soapy water. Some people may experience an allergic reaction and go into anaphylactic shock which can be lethal.

(Top left) Scorpion. (Bottom) Wasp, Tick

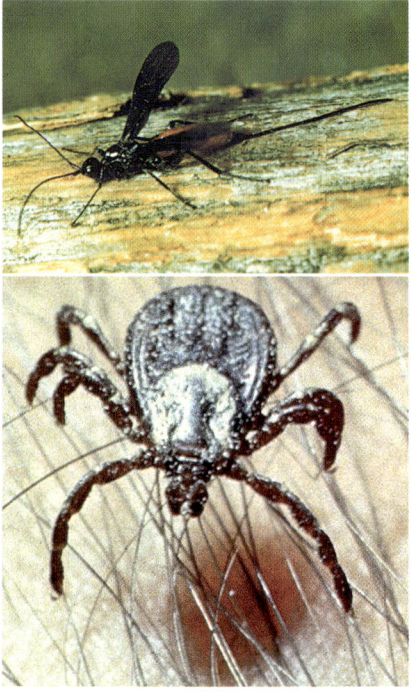

Ant and Yellowjacket

The Yellow Jackets and the Ants all lived together on the hillside about ten miles above Tsceminicum [LEWISTON, IDAHO] on the Clearwater River. The two families were quite friendly, although every once in a while members would get into an argument, which is no more than natural.

There was quite a bit of jealousy between the Chief of the Yellow Jackets and the Chief of the Ants. This was not real hatred, but each saw to it that his rights were not harmed. On the whole, the two bosses got along pretty well, considering their gossiping wives and their many children.

Chief Yellow Jacket was used to eating his meals on top of a certain rock, and he liked dried salmon the best. One day, he was seated on this rock, calmly eating a big dish of dried salmon which his wife had set before him.

Along came Chief Ant, and seeing Chief Yellow Jacket calmly eating his dinner, he became very angry. It is true that there were other rocks around for him to use, and he could have had dried salmon if he wished, but the sight of Chief Yellow Jacket made him very angry. "Hey there, you Yellow Jacket," he shouted at him, "What are you doing on the rock? I have as much right there as you. You can't eat there without asking me."

Chief Yellow Jacket looked up in surprise. "Why, Ant, what are you shouting about? I have always eaten my dinner on this rock."

"That makes no difference," said the Ant. "Why didn't you ask me about it?"

Yellow Jacket had by this time become very angry too. He rattled his wings and snapped his legs and yelled, "None of your business you little runt."

"Don't call me a runt," shouted Ant. "Nobody can insult me that way."

So saying that, Ant climbed up the side of the rock, and he and Yellow Jacket began to fight all over it. They fought face to face, and with arms locked about each other, they reared up on their hind legs, biting and poking for all they were worth.

Suddenly a great voice boomed out, "Here, you Ant and Yellow Jacket, stop that fighting."

It was Coyote, who happened to be passing down on the other side of the river. He had seen them struggling, but neither of them heard him because they were too busy fighting.

Again Coyote shouted, "You, Ant and you, Yellow Jacket, I order you to stop fighting. My subjects cannot fight. There is plenty of room and plenty of food for all of us, so why be foolish?"

This time they heard, but neither of them would stop.

A third time Coyote warned them, "This is the last time. I'm going to tell you now. Stop fighting or I shall turn you both into stone. You will no longer be great, for the La-te-tel-wit [HUMAN BEINGS] are coming.

They paid no heed, so Coyote used his magic medicine, waved his paws, and just as Ant and Yellow Jacket were arched together, Coyote turned them to stone.

To this day they remain for all to see, locked in each others arms on top of the big rock where Yellow Jacket ate his meals, but which became a battle ground because of greed.

dangerous creatures

(Above) Black widow. (Right) Velvet ant. (Facing page) Deer mouse.

Black Widow

Black widows are probably the most well-known and easily recognized species of venomous spiders. The adult female is jet black above with a red hour glass figure shape on the underside. Mature females grow to about one half inch in length. At Hanford, they like to inhabit rodent excavations and burrows and building sites. In healthy adults, black widow spider bites may cause painful muscle spasms for two to four days. Bitten individuals are often covered with perspiration; dizziness, nausea, and vomiting are common. Usually after 2-4 days all symptoms disappear.

Assassin Bug

Assassin bugs are generally brownish to black and are fairly common. At night they are attracted to light and they are predaceous on other insects. These bugs commonly enter households and can inflict a very painful bite.

Velvet Ant

These solitary wasps are so-called because they are antlike and are covered with a layer of dense fine hairs. The females are wingless and have a very painful sting. Many members of this family are very brightly colored.

Deer Mouse

Deer mice are found in all areas of the United States including the Hanford Site. Deer mice are 4-9 inches in length from head to tail, ranges in color from pale-gray to reddish-brown. They have white fur on their belly, feet, and underside of the tail. They can be a concern to man in that they are one of the principal carriers of the virus which causes the disease known as Hantavirus.

dangerous creatures

Hantavirus infection is a serious life-threatening illness caused by breathing in the virus. The virus is carried in the airborne particles of rodent urine, droppings, or saliva. The Hantavirus infection is often mistaken for bronchial pneumonia and symptoms are similar to the flu with body aches, chills, and troubled breathing. The four recommended ways to protect yourself from this disease are 1) Keep clean, wash dishes, and clean up spilled food 2) Prevent mice from entering 3) outside control: move woodpiles, gardens, and trash cans 100 ft from the home and 4) disinfect: always wear rubber gloves during clean up of rodent infested areas.

The qualities of nature enlighten my soul. In the quiet solitude of the forest, I find insight. I reflect on the glory of creation, the burden of my stewardship, my place in the world. The sanctity I see in nature comes from nature's effect on me.

I find sanctity in all of nature, however a cedar grove near my home embodies holiness. In this grove are towering old-growth cedars, reaching to heaven like pillars in a cathedral. There are luxurlant ferns, drawing the pure water flowing in a murmuring brook through the grove. Animals who have made the grove their home casually bask in the safety of the forest. The sacredness of the grove rejuvenates my soul, just as the moist, fragrant air rejuvenates my lungs.

When I venture into the cedar grove, life surrounds me—green life. The deep, dark green of cedar boughs; the light, youthful green of grasses; the soft, rich green of moss. If I listen hard enough, I can hear the grove speak to me—speaking not through the tongue of angels nor the word of man, but through the subtle greensong of nature. It is in nature where I feel closest to God, most in harmony with the earth, most at peace with myself. Nature is sacred; nature makes *me* sacred.
—Jeremy Crow, Nez Perce

Appendix 1:
Treaty with the Nez Percés, 1855.

Articles of agreement and convention made and concluded at the treaty ground, Camp Stevens, in the Walla - Walla Valley, this eleventh day of June, in the year one thousand eight hundred and fifty-five, by and between Isaac I Stevens, governor and superintendent of Indian affairs for the Territory of Washington, and Joel Palmer, superintendent of Indian affairs for Oregon Territory, on the part of the United States, and the undersigned chiefs, head-men, and delegates of the Nez Percé tribe of Indians occupying lands lying partly in Oregon and partly in Washington Territories, between the Cascade and Bitter Root Mountains, on behalf of, and acting for said tribe, and being duly authorized thereto by them, it being understood that Superintendent Isaac I. Stevens assumes to treat only with those of the above-named tribe of Indians residing within the Territory of Washington, and Superintendent Palmer with those residing exclusively in Oregon Territory.

Article 1. The said Nez Percé tribe of Indians hereby cede, relinquish, and convey to the United States all their right, title, and interest in and to the country occupied or claimed by them, bounded and described as follows, to wit: Commencing at the source of the Wo-na-ne-she or southern tributary of the Palouse River; thence down that river to the main Palouse; thence in a southerly direction to the Snake River, at the mouth of the Tucanon River; thence up the Tucanon to its source in the Blue Mountains; thence southerly along the ridge of the Blue Mountains; thence to a pint on the Grand Ronde River, midway between Grand Ronde and the mouth of the Woll-low-how River; thence along the divide between the waters of the Woll-low-how and Powder River; thence to the crossing of Snake River, at the mouth of Powder River; thence to the Salmon River, fifty miles above the place known [as] the "crossing of the Salmon River;" thence due north to the summit of the Bitter Root Mountains; thence along the crest of the Bitter Root Mountains to the place of beginning.

Article 2. There is, however, reserved from the lands above ceded for the use and occupations of the said tribe, and as a general reservation for other friendly tribes and bands of Indians in Washington Territory, not to exceed the present numbers of the Spokane, Walla-Walla, Cayuse, and Umatilla tribes and bands of Indians, the tract of land included within the following boundaries, to wit: Commencing where the Moh ha-na-she or southern tributary of the Palouse River flows from the spurs of the Bitter Root Mountains; thence down said tributary to the mouth of the Ti-nat-pan-up Creek; thence southerly to the crossing of the Snake River ten miles below the mouth of the Al-po-wa-wi River; thence to the source of the Al-po-wa-wi River in the Blue Mountains; thence along the crest of the Blue Mountains; thence to the crossing of the Grand Ronde River, midway between the Grand Ronde and the mouth of the Woll-low-how River; thence along the divide between the waters of the Woll-low-how and Powder Rivers; thence to the crossing of the Snake River fifteen miles below the mouth of the Powder River; thence to the Salmon River above the crossing; thence by the spurs of the Bitter Root Mountains to the place of beginning.

All which tract shall be set apart, and, so far as necessary, surveyed and marked out for the exclusive use and benefit of said tribe as an Indian reservation; nor shall any white man, excepting those in the employment of the Indian Department, be permitted to reside upon the said reservation without permission of the tribe and the superintendent and agent; and the said tribe agrees to remove to and settle upon the same within one year after the ratification of this treaty. In the mean time it shall be lawful for them to reside upon any ground not in the actual claim and occupation of citizens of the United States, and upon any ground claimed or occupied, if with the permission of the owner or claimant, guarantying, however, the right to all citizens of the United States to enter upon and occupy as settlers any lands not actually occupied and cultivated by said Indians at this time, and not included in the reservation above named. And provided that any substantial improvement heretofore made by any Indian, such as fields enclosed and cultivated, and houses erected upon the lands hereby ceded, and which he may be compelled to abandon in by said Indians at this time, and not included in the reservation above named. And provided that any substantial improvement heretofore made by any Indian, such as fields enclosed and cultivated, and houses erected upon the lands hereby ceded, and which may be compelled to abandon in consequence of this treaty, shall be valued under the direction of the President of the United States, and payment made for said Indian upon the reservation, and no Indian will be required to abandon the improvements aforesaid, now occupied by him, until their value in money or improvements of equal value shall be furnished him as aforesaid.

ARTICLE 3. And provided that, if necessary for the public convenience, roads may run through the said reservation, and, on the other hand, the right of way, with free access from the same to the nearest public highway, is secured to them, as also the right, in common with citizens of the Untied States, to travel upon all public highways. The use of the Clear Water and other streams flowing through the reservation is also secured to citizens of the United States for rafting purposes, and as public highways.

The exclusive right of taking fish in all the streams where running through or bordering said reservation is further secured to said Indians; as also the right of taking fish at all usual and accustomed places in common with citizens of the Territory; and of erecting temporary buildings for curing, together with the privilege of hunting, gathering roots and berries, and pasturing their horses and cattle upon open and unclaimed land.

ARTICLE 4. In consideration of the above cession, the United States agree to pay to the said tribe in addition to the goods and provisions distributed to them at the time of signing this treaty, the sum of two hundred thousand dollars, in the following manner, that is to say, sixty thousand dollars, to be expended under the direction of the President of the United States, the first year after the ratification of this treaty, in providing for their removal to the reserve, breaking up and fencing farms, building houses, supplying them with provisions and a suitable outfit, and for such other objects as he may deem necessary, and the remainder in annuities, as follows: for the first five years after the ratification of this treaty, ten thousand dollars each year, commencing September 1, 1856; for the next five years, six thousand each year, and for the next five years, four thousand dollars each year.

All which said sums of money shall be applied to the use and benefit of the said Indians, under the direction of the President of the United States, who may from time to time determine, at his discretion, upon what beneficial objects to expend the same for them. And the superintendent of Indian affairs, or other proper officer, shall each year inform the President of the wished of the Indians in relation thereto.

Article 5. The United States further agree to establish, at suitable points within said reservation, within one year after the ratification hereof, two schools, erecting the necessary buildings, keeping the same in repair, and providing them with furniture, books, and stationery, one of which shall be an agricultural and industrial school, to be located at the agency, and to be free to the children of said tribe, and to employ one superintendent of teaching and two teachers; to build two blacksmiths' shops, to one of which shall be attached a tinshop and to the other a gunsmith's shop; one carpenter's shop, one wagon and plough maker's shop, and to keep the same in repair, and furnished with the necessary tools and fixtures, and to employ two millers; to erect, keep in repair, and provide with the necessary furniture the buildings required for the accommodation of the said employees. The said buildings and establishments to be maintained and kept in repair as aforesaid, and the employees to be kept in service for the period of twenty years.

And in view of the fact that the head chief of the tribe is expected, and will be called upon, to perform many services of a public character, occupying much of his time, the United States further agrees to pay to the Nez Percé tribe five hundred dollars per year for the term of twenty years, after the ratification hereof, as a salary for such person as the tribe may select to be its head chief. To build for him, at a suitable point on the reservation, a comfortable house, and properly furnish the same, and to plough and fence for his use ten acres of land. The said salary to be paid to, and the said house to be occupied by, such head chief so long as he may be elected to that position by his tribe, and no longer.

And all the expenditures and expenses contemplated in this fifth article of this treaty shall be defrayed by the United States, and shall not be deducted from the annuities agreed to be paid to said tribe, nor shall the cost of transporting the goods for the annuity-payments be a charge upon the annuities, but shall be defrayed by the United States.

Article 6. The President may from time to time, at his discretion, cause the whole, or such portions of such reservation as he may think proper, to be surveyed into lots, and assign the same to such individuals or families of the said tribe as are willing to avail themselves of the privilege, and will locate on the same as a permanent home, on the same terms and subject to the same regulations as are provided in the sixth article of the treaty with the Omahas in the year 1854, so far as the same may be applicable.

Article 7. The annuities of the aforesaid tribe shall not be taken to pay the debts of individuals.

Article 8. The aforesaid tribe acknowledge their dependence upon the Government of the United States, and promise to be friendly with all citizens thereof, and pledge themselves to commit no depredations on the property of such citizens; and should any one or more of them violate this pledge, and the fact be satisfactorily proved before the agent, the property taken shall be returned, or in default thereof, or if injured or destroyed, compensation may be made by the Government out of the annuities. Nor will they make war on any other tribe except in self-defence, but will submit all matters of difference between them and the other Indians to the Government of the United States, or its agent, for decision, and abide thereby; and if any of the said Indians commit any depredations on any other Indians within the Territory of

Washington, the same rule shall prevail as that prescribed in this article in cases of depredations against citizens. And the said tribe agrees not to shelter or conceal offenders against the laws of the United States, but to deliver them up to the authorities for trial.

ARTICLE 9. The Nez Percés desire to exclude from their reservation the use of ardent spirits, and to prevent their people from drinking the same; and therefore it is provided that any Indian belonging to said tribe who is guilty of bringing liquor into said reservation, or who drinks liquor, may have his or her proportion of the annuities withheld from him or her for such time as the President may determine.

ARTICLE 10. The Nez Percé Indians having expressed in council a desire that William Craig should continue to live with them, he having uniformly shown himself their friend, it is further agreed that the tract of land now occupied by him, and described in his notice to the register and receiver of the land-office of the Territory of Washington, on the fourth day of June last, shall not be considered a part of the reservation provided for in this treaty, except that it shall be subject in common with the lands of the reservation to the operations of the intercourse act.

ARTICLE 11. The treaty shall be obligatory upon the contracting parties as soon as the same shall be ratified by the President and Senate of the United States.

In testimony whereof, the said Isaac I. Stevens, governor and superintendent of Indian affairs for the Territory of Washington, and Joel Palmer, superintendent of Indian affairs for Oregon Territory, and the chiefs, headmen, and delegates of the aforesaid Nez Percé of Indians, have hereunto set their hands and seals, at the place, and on the day and year hereinbefore written.

Bibliography

American Ornithologist's Union. 1983. *Checklist of North American Birds*. 6th ed. Washington D.C. American Ornithologist's Union.

———. 1989. *Thirty-seventh Supplement to the American Ornithologists Union Checklist of North American Birds*. Auk 106:532-538.

Aoki, H. 1994. *Nez Perce Dictionary*. Berkeley, California, University of California Press.

Aoki, H., and D. E. Walker Jr. 1989. *Nez Perce Oral Narratives, Linguistics Volume 104*. Berkeley, California: University of California Press.

Becker, J. M. 1993. *A Preliminary Survey of Selected Structures on the Hanford Site for Townsend's Big-eared Bat (Plecotus townsendii)*. PNL-8916. Richland, Washington, Pacific Northwest Laboratory.

Books, G. G. 1985. Avian Interactions with Mid-Columbia River Water Level Fluctuations. *Northwest Science* 59(4):304-312.

Burt, W. H., and R. P. Grossenheider. 1980. *Peterson Field Guide: Mammals*. 3rd ed. Boston, Massachusetts: Houghton Mifflin Company.

Cederholm, C. J., D. B. Houston and W. J. Scarlett. 1989. Fate of Coho Salmon (*Onycorhynchus kisutch*) Carcasses in Spawning Streams. *Can. J. Fish. Aquat. Sci.* Vol. 46.

Chatters, J. C. 1982. Prehistoric Settlement and Land Use in the Dry Columbia Basin. *Northwest Anthropological Research Notes*, Vol. 16, No. 2, pp. 125-147.

———. 1989. *Hanford Cultural Resources Management Plan*, PNL-6942. Richland, Washington: Pacific Northwest Laboratory.

Columbia River Inter-Tribal Fish Commission, 1996. *Wana Chinook Tymoo*. Issue one, 1996. Portland, Oregon.

Daubenmire, F. R. 1970. "Steppe Vegetation of Washington." *Washington Agricultural Experiment Station Technical Bulletin 62*. Pullman, Washington: Washington Agricultural Experiment Station.

Downs, J. L., W. H. Rickard, C. A. Brandt, L. L. Cadwell, C. E. Cushing, D. R. Geist, R. M. Mazaika, D. A. Neitzel, L. E. Rogers, M. R. Sackschewsky, and J. J. Nugent. 1993. *Habitat Types on the Hanford Site: Wildlife and Plant Species of Concern*. PNL-8942. Richland, Washington: Pacific Northwest Laboratory.

Ennor, H. R. 1991. *Birds of the Tri-Cities and Vicinity*. Richland, Washington: Lower Columbia Basin Audubon Society.

Fickeisen, D. H., R. E. Fitzner, R. H. Sauer, and J. L. Warren. 1980. *Wildlife Usage, Threatened and Endangered Species and Habitat Studies of the Hanford Reach, Columbia River, Washington*. Richland, Washington: Pacific Northwest Laboratory.

Fitzner, R. E. and K. R. Price. 1973. *The Use of Hanford Waste Ponds by Waterfowl and Other Birds*. BNWL-1738. Richland, Washington: Battelle Northwest Laboratory.

Fitzner, R. E., and R. H. Gray. 1991. The Status, Distribution and Ecology of Wildlife on the U.S. DOE Hanford Site: A Historical Overview of Research Activities. *Environmental Monitoring and Assessment* 18:173-202.

Fitzner, R. E. and W. H. Rickard. 1975. *Avifauna of Waste Ponds ERDA Hanford Reservation Benton County, Washington*. BNWL-1885. Richland, Washington: Battelle Northwest Laboratories.

Fitzner, R. E., W. H. Rickard, L. L. Cadwell, and L. E. Rogers. 1981. *Raptors of the Hanford Site and Nearby Areas of Southcentral Washington*. PNL-3212. Richland, Washington: Pacific Northwest Laboratory.

Gerber, M. S. 1992. *Legend and Legacy: Fifty Years of Defense Production at the Hanford Site*. WHC-MR-0293, Rev 1. Richland, Washington: Westinghouse Hanford Company.

———. 1993. *The Hanford Site: An Anthology of Early Histories*. WHC-MR-0435. Richland, Washington: Westinghouse Hanford Company.

Hanson, W. C. 1970. "Recent Sight Records of Jaegers in Southeastern Washington." *Murrelet*, 51:17.

Herde, K. E. 1952. *Logbook for Biological Monitoring*. 11W-2813-1. Richland, Washington: Hanford Atomic Productions Operations, General Electric.

Hunn, E. S. 1990 *Nch'i-Wana 'The Big River' Mid-Columbia Indians and Their Land*. Seattle, Washington: University of Washington Press.

Josephy, A. M. 1979 *The Nez Perce Indians and the Opening of the Northwest*. Lincoln, Nebraska: University of Nebraska Press.

Landeen, D. S., A. R. Johnson, and R. M. Mitchell. 1992. *Status of Birds at the Hanford Site in Southeastern Washington*. WHC-EP-0402 Rev. 1. Richland, Washington: Westinghouse Hanford Company.

Lopez, B. 1977. *Giving Birth to Thunder Sleeping With His Daughter Coyote Builds North America*. New York, New York: Avon Books.

McLulan, T. C. 1971. *Touch the Earth, A Self-Portrait of Indian Existence*. New York, New York: Promontory Press.

National Geographic Society. 1989. *Field Guide to the Birds of North America*. 2nd ed. Washington, D.C.: National Geographic Society.

Nature Conservancy. 1995. *Biodiversity Inventory and Analysis of the Hanford Site 1995 Annual Report*. J.A. Soll and C. Soper, eds. U.S. DOE Grant Award Number DE-FG06-94rl12858. Seattle, Washington: The Nature Conservancy.

———. 1994. *Biodiversity Inventory and Analysis of the Hanford Site 1994 Annual Report*. Seattle, Washington: Nature Conservancy.

Olson, D. L. 1995. *Shared Spirits: Wildlife and Native Americans*. Minocqua, Wisconsin: NorthWord Press Inc.

Osborne, D. H. 1957. *Excavations in the McNary Reservoir near Umatilla, Oregon*. Washington D.C.: Bureau of American Ethnology, Bulletin 166.

Pacific Northwest Laboratory, 1995. *Prehistoric Period Context Statement for the U.S. Department of Energy's Hanford Site Benton County*, Washington D.C.: Draft Document.

Rickard, W. H. 1964. "A Vagrant Occurrence of the Black Phoebe in Southeastern Washington." *Condor*, vol. 66.

Rickard, W. H., J. D. Hedlund, and R. G. Shreckhise. 1974. *Mammals of the Hanford Reservation in Relation to Management of Radioactive Waste*. BNWL-1877. Richland, Washington: Pacific Northwest Laboratory.

Sackschewsky, M. R., G. I. Baird, D. S. Landeen, J. L. Downs, and W. H. Rickard. 1992. *Vascular Plants of the Hanford Site*. WHC-EP-0554. Richland, Washington: Westinghouse Hanford Company.

Schalk, R. F. and D. Olson. 1983. "The Faunal Assemblage." In *The 1978 and 1979 Excavations at Strawberry Island in the McNary Reservoir*. Edited by R.S. Schalk. Project Report No. 19. Laboratory of Archaeology and History, Pullman, Washington: Washington State University and History.

Schroedl, G. F. 1973. *The Archeological Occurrence of Bison in the Southern Plateau*. Laboratory of Anthropology Reports of Investigations, No. 51. Pullman, Washington: Washington State University

Shawley, S. D. 1974. *Nez Perce Dress: A Study in Culture Change*. Moscow, Idaho: University of Idaho Department of Sociology/Anthropology.

Slickpoo, A. P. Sr., and D. E. Walker. 1973. *Noon Nee-Me-Poo (We, The Nez Perces)*.

Spinden, H. J. 1908. *Memoirs of the American Anthropological Association, The Nez Perce Indians, Volume II Part 3*. Lancaster, Pennsylvania: New Era Printing Company.

Stegen, J. A. 1992. *Biological Assessment for State Candidate and Monitor Wildlife Species Related to CERCLA*. WHC-SD-EN-EE-009. Richland, Washington: Westinghouse Hanford Company.

Stepniewski, A. M. 1994. *Birds of the Wahluke Slope (Saddle Mountain NWR/Wahluke Slope Wildlife Area), Hanford Site Biodiversity Inventory*. Report to the Nature Conservancy Contract #WAFO-022094.

Stone, J., ed. 1993. *Every Part of This Earth is Sacred, Native American Voices in Praise of Nature*. San Francisco, California: Harper Collins Publishers.

Stone, W. A., J. M. Thorp, O. P. GiVord, and D. J. Hoitink. 1983. *Climatological Summary for the Hanford Area*. PNL-4622. Richland, Washington: Pacific Northwest Laboratory

Taylor, C. F., ed. 1994. *Native American Myths and Legends*. New York, New York: Smithmark Publishing Inc.

Walker, D. E. 1994. *Blood of the Monster The Nez Perce Coyote Cycle*. Worland, Wyoming: High Plains Publishing Company.

Washington Department of Fish and Wildlife. 1995. *Priority Habitats and Species List: Habitat Program*. Olympia, Washington: Department of Fish and Wildlife.

———. 1995. *Washington State Management Plan for Sage Grouse*. Olympia, Washington: Game Div., Wash. Dept. Fish and Wildl.

Washington Department of Wildlife. 1993. *Rare Bats of the Shrub-Steppe Ecosystem of Eastern Washington*. Olympia, Washington: Washington Department of Wildlife, Nongame Program.

———. 1993. *Status of the Pygmy rabbit* (Brachylagus idahoensis) *in Washington*. Olympia, Washington: Wash. Dept. Wildl.

Washington Natural Heritage Program. 1994. *Endangered, Threatened, and Sensitive Vascular Plants of Washington*. Olympia, Washington: Department of Natural Resources.

Weber, J. W. and E. J. Larrison. 1977. *Birds of Southeastern Washington*. Moscow, Idaho: University Press of Idaho, University Station.

ACKNOWLEDGMENTS

Special thanks to the many people and organizations who provided the photos that were used in this publication, many of which were taken at the Hanford Site and in the Columbia Basin region. Thanks to the following people who provided information and review comments: Horace Axtell, Liz Block, Elmer Crow Jr., Lynda Crow, Kim Cunningham, Freida Ellenwood, Joe Fitch, Bob Grimm, Jim Hepworth, Carla HighEagle, Ray Johnson, Julie Kane, Keith Lawrence, Jay McConnaughey, Ron Mitchell, Sam Penney, Ken Petersen, Josiah Pinkham, Donna Powaukee, Julie Simpson, Tony Sitner, Allen Slickpoo Sr., Mae Taylor, Jerry Todd, Lynda Williams and the Nez Perce Tribe Executive Committee. Photo credits are listed below.

American Museum of Natural History: short horned lizard (10), mallard nest (35), pallid bat (60), brown recluse (82), rattlesnake (83), wasp (84), tick (84), yellow jacket (85), black widow (86), velvet ant (86), deer mouse (87)

American Society of Mammalogy: otter (9), little brown myotis (60), silver-haired bat (60), short-tailed weasel (63), northern grasshopper mouse (68), sagebrush vole (68)

Don Baccus (email donb@rational.com): Golden eagle (11), sora (40), American avocet (TITLE, 41), black-necked stilt (42), common snipe (43), least sandpiper (43), western sandpiper (43), Wilson's phalarope (43), short-eared owl (45), common nighthawk (46), northern flicker (48), eastern kingbird (49), sage thrasher (54), white-crowned sparrow (56), lark sparrow (56), Brewer's blackbird (55), brown-headed cowbird (57)

Boeing Computer Services: lupine (2-3), butterfly on penstemon (8), Hanford Reach (8), mule deer (8), coyote (9), blazing star (10), mariposa lily (12), rattlesnake (14), chinook salmon (16), reactor and geese (21), Hanford map (21), onion (22), Rattlesnake Mountain (23), mule deer (24-25), white pelican (31), ferruginous hawk (37), sage grouse (38), butterflies (39), evening primrose (39), long-billed curlew (42), coyote (COVER, 64), cottontail rabbit (71), side-blotched lizard (74-75), scorpion (84)

Corel Film Services: ducks (4-5), butterfly (6-7), mountain sheep (26), avocet (28-29), bittern (32), snowy egret (32), black-crowned night heron (32), Great blue heron (32), osprey (33), Canada geese (34), wood duck (34), bufflehead (35), common goldeneye (35), American bald eagle (36), peregrine falcon (37), Sandhill cranes (40), Great horned owl (45), hummingbird sp. (47), marsh wren (52), bison (58-59), cougar (61), raccoon (61), badger (63), otter (63), skunk (63), bobcat (64), beaver (66, 67), marmot (66), muskrat (68), mule deer (72), elk (73)

Howard Ennor: Western grebes (30), Great blue heron (32), northern shoveler (34), barn owl (45), western kingbird (49), western kingbird nest (49), mountain bluebird (53), American robin (53), loggerhead shrike (54), rufous-sided towhee (55), evening grosbeak (57), red-winged blackbird (57), yellow headed blackbird (COVER, 57)

Eternal Flame Photography: waterfall (15), tree stump (87), Ed Crow (91)

Carla HighEagle: Apaloosa horse (19)

Dan Landeen: desert parsley (12), chinook salmon carcass (15), Columbia River (17), shooting star (20), Umtanum ridge (20), lichen (21), Columbia River (22), Mae Taylor (27), rainbow (44), West Lake (46), sage sparrow (49, 56), badger den (63), pocket gopher (67), pocket mouse mound (68), Umtanum ridge (53), prickly pear cactus (54), sage sparrow (56), Great Basin pocket mouse (67), pocket gopher cast (67), hunting blind (71), gopher snake (77), fishing spider (80-81), cactus (83)

Reid Landeen: spotted sandpiper tracks in mud (94)

Ed Miller: horned grebe (30), great egret (32), osprey (37), American kestrel (37), greater yellowlegs (43), canyon wren (52), horned lark nest (50), bohemian waxwing (54)

Nature Conservancy: Hanford Reach (20), sand dunes (21), sagebrush lizard (77), desert night snake (77), Woodhouse's toad (79)

Nez Perce National Historical Park: swan head carving (11), headdress (11), Lewis and Clark journal (12), Nez Perce woman (12), Nez Perce man (12), war whistle (13), deer antler helmet (13), bone deflesher (13), necklace (13), rattles (14), Indian fishing (15), Celilo Falls (16), sucker bone (17), petroglyph (18), Chief Joseph (18), woven grass hat (19)

Clyde Pritchett: bushy-tailed woodrat (68)

John B. Ridley Research Library, Quetico Park: common loon (30), Bonaparte's gulls (41), chipping sparrow (56),

Ron Spomer: eared grebe (30), double-crested cormorants (31), hooded merganser (34), American wigeon in flight (34), ruddy duck (35), black-billed magpie (51), lazuli bunting (55)

Jerry Todd: Herman Reuben (5)

United States Fish and Wildlife Service: butterfly (17, 20), trumpeter swans (35), American coot on nest (40), killdeer (41), killdeer nest (41), ring-billed gull (42), semi-palmated plover (42), rock doves (44), mourning dove (44), burrowing owls (45), long-eared owl (45), horned lark (50), all swallows sp. (50), Townsend ground squirrel (66), western skink (77), Pacific tree frog (79), Great Basin spadefoot toad (79), porcupine (69), white-tailed jackrabbit (70), gopher snake (76), side-blotched lizard (77), bullfrog (79) green racer (77), painted turtle (78), leopard frog (78), meadow lark (92)

Maurice Vial: prairie falcon (36), white-throated swift (46), Lewis' woodpecker (48), downy woodpecker (48), rock wren (53), black-throated sparrow (56)

John Wallace: Stellar's jay (51)

Wyoming Game and Fish: Hungarian partidges (38), long-tailed weasel (63), short-horned lizard (76)

Yakama Indian Nation: western meadowlark (11, 57), Hanford elk (72)

Hanford Wildlife Checklist

Birds

- Avocet, American (41)
- Bittern
  - American (32)
- Blackbird
  - Brewer's (57)
  - Red-winged (57)
  - Rusty (57)
  - Yellow-headed (57)
- Bluebird
  - Mountain (53)
  - Western (53)
- Bobolink (57)
- Bobwhite, Northern (38)
- Brant (35)
- Bufflehead (35)
- Bunting
  - Lazuli (55)
  - Snow (57)
- Canvasback (35)
- Catbird, Gray (54)
- Chat, Yellow-breasted (55)
- Chickadee, Black-capped (52)
- Chukar (38)
- Coot, American (40)
- Cormorant, Double-crested (31)
- Cowbird, Brown-headed (57)
- Crane, Sandhill (40)
- Creeper, Brown (52)
- Crow, American (51)
- Curlew, Long-billed (43)
- Dove
  - Mourning (44)
  - Rock (44)
- Dowitcher
  - Long-billed (43)
  - Short-billed (43)
- Duck
  - Ring-necked (35)
  - Ruddy (34)
  - Wood (35)
- Dunlin (43)
- Eagle
  - Bald (36, 37)
  - Golden (37)
- Egret
  - Great (32)
  - Snowy (32)
- Falcon
  - Peregrine (37)
  - Prairie (36)
- Finch
  - Cassin's (57)
  - House (57)
  - Purple (57)
  - Rosy (57)
- Flicker, Northern (44)
- Flycatcher
  - Ash-throated (49)
  - Cordilleran (49)
  - Dusky (49)
  - Gray (49)
  - Hammond's (49)
  - Olive-sided (49)
  - Pacific-slope (49)
  - Willow (49)
- Gadwall (34)
- Godwit, Marbled (43)5
- Goldeneye
  - Barrow's (35)
  - Common (35)
- Goldfinch
  - American (57)
  - Lesser (57)
- Goose
  - Aleutian Canada (34)
  - Canada (34)
  - Greater White-fronted (35)
  - Snow (35)
- Goshawk, Northern (37)
- Grebe
  - Clark's (30)
  - Eared (30)
  - Horned (30)
  - Pied-billed (30)
  - Red-necked (30)
  - Western (30)
- Grosbeak
  - Black-headed (55)
  - Evening (57)
  - Rose-breasted (55)
- Gull
  - Bonaparte's (41,42)
  - California (42)
  - Franklin's (42)
  - Glaucous-winged (42)
  - Herring (42)
  - Ring-billed (42)
  - Sabine's (42)
- Gyrfalcon (37)
- Harrier, Northern (36)
- Hawk
  - Cooper's (37)
  - Ferruginous (36)
  - Red-tailed (36)
  - Rough-legged (37)
  - Sharp-shinned (37)
  - Swainson's (36)
- Heron
  - Great Blue (32)
- Hummingbird
  - Anna's (46)
  - Black-chinned (46)
  - Calliope (46)
  - Rufous (46)
- Jaeger
  - Long-tailed (42)
  - Parasitic (42)
- Jay
  - Scrub (51)
  - Stellar's (51,
- Junco, Dark-eyed (55)
- Kestrel, American (36,37)
- Killdeer (41)
- Kingbird
  - Eastern (49)
  - Western (49)
- Kingfisher, Belted (47)
- Kinglet
  - Golden-crowned (53)
  - Ruby-crowned (53)
- Kittiwake
  - Blacklegged (42)
- Lark, Horned (50)
- Longspur, Lapland (55)
- Loon
  - Common (30)
  - Pacific (30)
  - Red-throated (30)
- Magpie
  - Black-billed (51)
- Mallard (34)
- Meadowlark, Western (57,
- Merganser
  - Common (35)
  - Hooded (35)
  - Red-breasted (35)
- Merlin (37)
- Mockingbird, Northern (54)
- Night-Heron, black-crowned(32
- Nighthawk, common (46)
- Nutcracker, Clark's (51)
- Nuthatch, Red-breasted (52)
- Oldsquaw (35)
- Oriole, Northern (57)
- Osprey (37)
- Owl
  - Barn (45)
  - Barred (45)
  - Burrowing (45)
  - Flammulated (45)
  - Great Horned (45)
  - Long-eared (45)
  - Nothern saw-whet (45)
  - Short eared (15)
  - Snowy (45)
  - Western Screech (45)
- Partridge, Gray (38)
- Pelican, American White (31)
- Pewee, Western Wood (49)
- Phalarope
  - Red (43)
  - Wilson's (43)
- Pheasant, Ring-necked (38)
- Phoebe
  - Black (49)
  - Say's (49)
- Pigeon, Band-tailed (44)
- Pintail, Northern (35)
- Pipit, American (54)
- Plover
  - Black-bellied (41)
  - Semipalmated (41)
  - Mountain (41)
- Poorwill, Common (46)

95

- ❏ Quail, California (38)
- ❏ Rail, Virginia (40)
- ❏ Raven, Common (51)
- ❏ Redhead (34)
- ❏ Redstart, American (55)
- ❏ Robin, American (53)
- ❏ Sanderling (43)
- Sandpiper
  - ❏ Baird's (43)
  - ❏ Least (43)
  - ❏ Pectoral (43)
  - ❏ Semipalmated (43)
  - ❏ Sharp-tailed (43)
  - ❏ Solitary (43)
  - ❏ Spotted (43)
  - ❏ Western (43)
- Scaup
  - ❏ Greater (35)
  - ❏ Lesser (35)
- ❏ Shoveler, Northern 34
- Shrike
  - ❏ Loggerhead (54)
  - ❏ Northern (54)
- ❏ Siskin, Pine (57)
- ❏ Snipe, Common (43)
- ❏ Solitaire, Townsend's (53)
- ❏ Sora (40)
- Sparrow
  - ❏ American Tree (55)
  - ❏ Black-throated (55)
  - ❏ Brewer's (55)
  - ❏ Chipping (55)
  - ❏ Fox (55)
  - ❏ Golden-crowned (55)
  - ❏ Grasshopper (56)
  - ❏ Harris' (55)
  - ❏ House (57)
  - ❏ Lark (55)
  - ❏ Lincoln's (55)
  - ❏ Sage (55,56)
  - ❏ Savannah (55)
  - ❏ Song (55)
  - ❏ Swamp (55)
  - ❏ Vesper (55)
  - ❏ White-crowned (56)
- ❏ Starling, European (54)
- ❏ Stilt, Black-necked (41)
- Swallow
  - ❏ Bank (50)
  - ❏ Barn (50)
  - ❏ Cliff (50)
  - ❏ Northern rough-winged (50)
  - ❏ Tree (50)
  - ❏ Violet Green (50)
- Swan
  - ❏ Trumpeter (34)
  - ❏ Tundra (34)
- ❏ Swift, White-throated (46)
- ❏ Tanager, Western (55)
- Teal
  - ❏ Blue-winged (34)
  - ❏ Cinnamon (34)
  - ❏ Green-winged (34)
- Tern
  - ❏ Arctic (42)
  - ❏ Black (42)
  - ❏ Caspian (42)
  - ❏ Common (42)
  - ❏ Forster's (42)
- ❏ Thrasher, Sage (54)
- Thrush
  - ❏ Hermit (53)
  - ❏ Varied (53)
- ❏ Towhee, Rufous-sided (55)
- Vireo
  - ❏ Hutton's (55)
  - ❏ Philadelphia (55)
  - ❏ Red-eyed (55)
  - ❏ Solitary (55)
  - ❏ Warbling (55)
- ❏ Vulture, Turkey (37)
- Warbler
  - ❏ Blackpoll (55)
  - ❏ MacGillivray's (55)
  - ❏ Nashville (55)
  - ❏ Orange-crowned (55)
  - ❏ Palm (55)
  - ❏ Tennessee (55)
  - ❏ Townsend's (55)
  - ❏ Wilson's (55)
  - ❏ Yellow (55)
  - ❏ Yellow-rumped (55)
- Waxwing
  - ❏ Bohemian (54)
  - ❏ Cedar (54)
- Wigeon
  - ❏ American (35)
  - ❏ Eurasian (35)
- ❏ Willet (43)
- Woodpecker
  - ❏ Downy (48)
  - ❏ Hairy (48)
  - ❏ Lewis' (48)
- Wren
  - ❏ Bewick's (53)
  - ❏ Canyon (52)
  - ❏ House (53)
  - ❏ Marsh (52)
  - ❏ Rock (53)
  - ❏ Winter (53)
- Yellowlegs
  - ❏ Greater (43)
  - ❏ Lesser (43)
- ❏ Yellowthroat, Common (55)

Mammals

- ❏ Antelope, Pronghorn (4,13)
- ❏ Badger (63)
- Bats
  - ❏ Little brown myotis (60)
  - ❏ Pallid (60)
  - ❏ Silver-haired (60)
  - ❏ Small-footed myotis (60)
  - ❏ Western pipistrel (60)
  - ❏ Yuma myotis (60)
- ❏ Beaver (66,67)
- Cats
  - ❏ Bobcat (64)
  - ❏ Cougar (61,64)
- ❏ Chipmunk, Least (66)
- ❏ Coyote (64)
- Deer
  - ❏ Mule deer (72)
  - ❏ White-tailed (72)
- ❏ Elk (72,73)
- Ground Squirrel
  - ❏ Townsend (66)
- Mouse
  - ❏ Deer (68,87)
  - ❏ Harvest (68)
  - ❏ House (68)
  - ❏ Northern Grasshopper (68)
- Jackrabbit
  - ❏ Blacktail (77)
  - ❏ Whitetail (77)
- Marmot
  - ❏ Yellowbelly (66)
- ❏ Mink (63)
- ❏ Muskrat (68)
- ❏ Otter (63)
- Pocket Gopher
  - ❏ Northern (67)
- Pocket mouse
  - ❏ Great Basin (67)
- ❏ Porcupine (69)
- Rabbit
  - ❏ Cottontail (70,71)
- ❏ Raccoon (61)
- Rat
  - ❏ Norway (68)
- Shrew
  - ❏ Merriam's (60)
  - ❏ Vagrant (60)
- ❏ Skunk, Striped (63)
- Vole
  - ❏ Montane (68)
  - ❏ Sagebrush (68)
- Weasel
  - ❏ Long-tailed (63)
  - ❏ Short-tailed (63)
- Woodrat
  - ❏ Bushy-tailed (68)

REPTILES AND AMPHIBIANS

- Frog
  - ❏ Pacific Tree (79)
  - ❏ Leopard (78)
  - ❏ Bullfrog (79)
- Skink
  - ❏ Western Skink (78)
- Snake
  - ❏ Green Racer (77)
  - ❏ Rattlesnake (77,83)
  - ❏ Striped Whipsnake (76)
  - ❏ Gopher (77)
  - ❏ Desert Night (77)
- Lizard
  - ❏ Sagebrush (77)
  - ❏ Short-horned (10,76,77)
  - ❏ Side-blotched (77)
- Toad
  - ❏ Great Basin Spadefoot (79)
  - ❏ Woodhouse's (79)
  - ❏ Western (78)
- Turtle
  - ❏ Painted (79)